SALVAGE

THE UNDERWATER WORLD

OTHER BOOKS BY JIM THORNE

The Underwater World

A Survey of Oceanography Today

JIM THORNE

Thomas Y. Crowell Company

New York   Established 1834

Photographs, unless otherwise credited, were taken by the author.

Copyright © 1969 by Jim Thorne

Manufactured in the United States of America

L.C. Card 69-18673

2 3 4 5 6 7 8 9 10

Acknowledgments

With special thanks to the following for their contributions to the writing of this book or for their contributions to oceanography in general:

Warrant Officer Robert Barth, U.S. Navy
Edward H. Heinemann, General Dynamics Corporation
James Duggan
Dr. John C. Lilly
Margaret Howe
D. C. Pauli, U.S. Office of Naval Research
Parliamentary Group for World Government, London
Admiral T. B. Owen, Chief of Naval Research, U.S. Navy
Lieutenant Commander C. W. Larson, Public Affairs Officer
Ivan Tors
Electric Boat Division of General Dynamics Corporation
Ocean Systems, Inc.
Corning Glass Works
Ric O'Feldman
Don Renn
EG&G International
George Garmus, Basford, Inc.
Ken Wilsey, Jordan Associates
Ogden Technological Laboratories
Bendix Corporation
Sir John Davis
Captain G. F. Bond
Captain Walter Mazzone
John L. Mero, Ocean Resources, Inc.
Sir John Foster, K.B.E., Q.C.M.P., London
Patrick Armstrong, Honorable Secretary, Parliament for World Government of the Sea's Resources

Contents

Illustrations

THE UNDERWATER WORLD

The author sits atop the multipurpose submarine *Star III*. Operating at depths of 2,000 feet and speeds of six knots, *Star III* is well suited to underwater mapping and search and recovery work. (*General Dynamics, Electric Boat Division*)

1 · *It's a Small World*

Is it?

It's a familiar cliché and I'm certain that we have all uttered it at one time or another. Usually we followed that statement with another one having to do with the relativity of time and space—the jet age, the shrinking world, and so on. We believed that, too, when we said it, but nothing could be farther from the truth! Let's take a close look at a few facts and bury that trite "it's a small world" phrase forever.

A few years ago the United States Navy plunged its two-man bathyscaphe *Trieste* seven miles down into the black, secret depths of the Marianas Trench in the Pacific, reaching the greatest ocean depth ever recorded, 37,800 feet.

Man has also conquered the highest point on earth, Mt. Everest in Tibet. This peak reaches 29,002 feet—not quite six miles—above sea level.

Six miles up and seven miles down. Not bad! But the diameter of the earth is 7,900 miles! There are approximately 191 million square miles of earth's surface, of which 147 million square miles are water. With this in mind, we can compare man's accomplishments to those of a child carving his initials on the rock of Gibraltar with a safety pin!

In all truth, however, we must realize that this is a "small world" in a few ways. In travel, the jet age *has* done quite a bit to shrink the world's miles. Economically, the world is rapidly getting too "small" for its population.

The world population today is approximately 2.7 billion and growing at a frightening rate. By the year 2000 it is expected to reach 6.27 billion, at which time there may not be enough food to go around.

Two thousand calories per day for each person is considered minimum for normal health. If human food consumption drops below this, the results are lethargy, reduced productivity, and ill health. Today, in areas such as India, Africa, Japan, China, and Peru, the populace consumes much less than the minimum requirement in calories and hardly enough animal protein to exist at all. If the sea's food potential is developed to its fullest it can do much toward mitigating this world hunger problem, especially in the production of high-protein concentrates from plankton and whole fish. Although work in this area has begun, we are still a long way from harvesting the seas with the facility that farmers work the land.

The total plant life of the sea comprises ninety percent of the entire vegetable growth of the world. Basically, plankton, which is the grazing fodder of the sea, is of two types: phytoplankton, minute *plants* using energy from sunlight to extract and utilize minerals dissolved in seawater; and zooplankton, minute *animals* that graze on phytoplankton and subsequently furnish food for the sea's fish population. Even some of the sea's largest mammals, such as whales, live on plankton.

Two possibilities can result from an increase of plankton: a vastly increased fish population and the definite possibility of processing plankton—which, like fish, is high in protein—into a nutritious and palatable human food. It may be some time before this item is listed on the menu at Sardi's, but the day is coming when it will certainly appear on the home menu. It will probably take the form of a concentrate which can be added to other foods, although recent experiments in Japan have succeeded in taking out the "fishy" taste, making it possible to serve plankton as a dinner dish. Fish and plankton byproducts are also valuable as a feed supplement for chicken, hogs, sheep, and cattle.

The best and most efficient method of increasing plankton would be to create, mechanically, an upwelling of the deeper layers of water, which are rich in nutrients, and bring the plankton to the surface. Here, at the surface where sunlight and nutrients could work their natural miracle, zones would be created which would become, in effect, plankton "farms" ready for harvest.

A natural result of these "sea farms" would be a vastly increased fish population, and this, too, would do much toward solving the food problem.

Thinking in terms of stimulating the plankton population to increase fish population, let's consider these statistics: The production of one pound of cod requires 10 pounds of small fish, which require 100 pounds of other fish (any and all types), which require 1,000 pounds of crustaceans, which require 10,000 pounds of zooplankton, which require 100,000 pounds of phytoplankton.

This food chain equation shows clearly the need to create vast areas of plankton farms in order to increase the world fish population. Although the formula is relatively simple the task of accomplishment is prodigious and requires a great deal of time. The payoff, however, may be quite large because of the high nutritional value of fish. One ton of fish, for example, is equal in food value to the meat yield of thirty-five to forty head of cattle on the range. It requires this many animals on the hoof in order to send four or five prime cattle to the slaughterhouse. Therefore, a country that has the opportunity of catching 100,000 tons of fish per year has the same potential food wealth as a country that has 3.5 to 4 million head of cattle on the range. And improvements in technology are found to increase the fishing potential of seafaring nations. In the foreseeable future, fishing will no longer be limited in certain regions by seasonal ice areas or high waves. In past books I have described expeditions in which we have tested the Aqua-Therm, a submersible pump which can "melt" even Antarctic ice and can also flatten high waves. This device has begun to be installed in certain lakes and rivers to keep them free of ice during the winter months, but it has an even greater potential.

The present estimate is that, after enrichment (photosynthesis), there are ten pounds of collectable phytoplankton per thousand tons of seawater. It has been proposed that the upwelling of the deeper waters (bringing phytoplankton near the surface) be accomplished by heating the water with nuclear reactors. The cost alone makes this system impractical. This kind of nuclear thermal pumping would cost between two and three dollars per pound of plankton.

By using the Aqua-Therm, however, the cost of raising plankton could be less than one cent per pound. This artificial raising of deep-water nutrients would result in vastly increased plankton "pastures," which would also multiply the fish population by a factor of several hundred.

Simply stated, here is the way it would work: A propeller of special configuration activated by a submersible pump or motor (the Aqua-Therm) can drive a column of water from almost any depth to the surface. In effect, it creates a vertical current that could be used to surface the nutrients necessary for the production of plankton. Should it be discovered that aeration of the water is beneficial, a simple modification would aerate the water with a rising curtain of countless small bubbles. Plankton farms would thus be created over a wide expanse of ocean, providing vast pastures for fish. This would unquestionably increase fish population as well as the protein potential from plankton itself.

An unfortunate situation exists relative to the potential of "inner space." Our federal government and its policy makers are so concerned with the political importance of being the first on the arid, economically valueless moon, that they have paid scant attention to the more accessible and economically more vital "inner space" or waters of the world. Annual monies budgeted for the development of underwater resources are far less then those allocated to outer space or our "advisory wars" in such places as Korea or Vietnam.

Some governmental departments, however, are accomplishing as much as possible within their limited budgets, and they deserve the highest praise for their efforts.

The agencies concerned with fisheries, for example, have increased the tonnage of our fishing boats, developed more efficient methods for locating large schools of fish, and improved methods of catch. Still, we are behind Peru, Japan, China, and Russia as far as annual fish catch is concerned. This condition should not exist. We are bordered by two of the world's greatest oceans. They contain enough fish to bolster the wealth of this country. We can acquire it by further development of our fishing fleets.

The U. S. Navy has accomplished a great deal toward proving that man can perform useful work in the still hostile atmosphere of deep water. This was done through a series of underwater habitat experiments called Sealab, which will be discussed at length later in this book. These experiments took men to a depth of 630 feet, where they lived and worked for a period of forty-five days.

Private industry, too, has become increasingly interested in oceanography. Farseeing projects and experiments are being conducted daily. Scientists and engineers are exploring a wide spectrum of the sea's enormous potential. The largest portion of activity, however, takes place in two areas: the design and development of submersibles, especially submarines for deep diving work, and the exploitation of offshore petroleum products.

The study of oceanography today has infinite variety. In this book we will explore all the "happenings" currently taking place in the underwater world and see how they affect our economic life. We will take a close look at such things as the development of submarines, underwater photography, sport diving, treasure hunting, archaeology, salvage, mining, drilling for oil, fishing, and man's latest experiments in living underwater.

By exploring the "inner space," man is—in a sense—returning to his birthplace, for the oceans are the source of all life.

The planet earth was formed somewhere between four and six billion years ago. At the very beginning there may have been a heavy atmosphere of thick, black clouds extending upward for miles and blotting out the sun.

As the earth began to cool, condensation occurred and rain began to fall—slowly at first, then faster and faster until it became a deluge, which continued for an unknown period of time. This solid sheet of rain drove down as though there were no end, until the clouds had poured approximately 300 million cubic miles of water on the bare rock—for that was all there was on the earth at that time, only rock. Not a single grain of sand, nor a piece of dirt, nor a blade of grass, nor any living thing!

The rain had done its job. The deep cracks and basins were brimming over with fresh water, clean and clear.

In most areas (the flat lands) a complete quiet pervaded—the calm of inland lakes. Even the huge oceans were probably like millponds. In other places, gravity began to cause waters to make the first carving, eroding movements that eventually produced rivers.

At the same time another process started in the great ocean basins —the interminable process of washing salts and minerals out of the rocks into the waters of the sea.

There was still no life of any kind. Some theorists say life began to appear about two billion years ago. Modern science can only assume that the first living things must have been molecules that had the ability to reproduce themselves. It would appear that this must first have happened at sea, because water was indispensable to their formation.

It is possible that something like the following circumstances caused their formation: a blanket of clouds, warm and incubating, and an atmosphere of marsh gas and ammonia creating an oceanic cauldron of dissolving minerals in which large molecules of carbon combined and recombined until the correct combination was finally reached, leading to the first simple, single-celled organism.

After this miracle, something else happened which was very important to the development of life. Certain living organisms acquired the ability to capture energy from the sunlight and used it to make food (nourishment) out of the chemicals contained in the sea. They took carbon dioxide dissolved in the sea, and by using sunlight were able to produce sugar and, at the same time, release oxygen. These organisms were, in effect, plants.

Other organisms, unable to perform this miracle (photosynthesis), but still needing organic food (living organisms) for survival, found they could live by devouring the plants. These were the first animals.

The speculation is hardly flattering to the human ego, but it is probable that man evolved from a jellyfish, for the lowly jellyfish was the first creature to possess both mouth and stomach.

Next, in this nursery of life, there developed animals with protective shells. During the following era the same animals found a way to get out of their shells, and in many cases trailed these shells as far as sixteen feet behind them.

During a later period known as the Silurian Age (about 360 million years ago), a small ugly sea scorpion crawled out for a day on the beach and, managing to adapt himself, decided to stay.

It was during the next period, the Devonian (about 330 million years ago) that a few fish left the sea and followed the scorpion, who was already beginning to do pretty well. They developed into the first land-dwelling vertebrates. According to biological theory, man is descended from one of these adventurous land-invading fish.

Slowly populations grew. The land and sea teemed with life. But the beautiful sea held the final fascination, for the former creatures of the sea began to return. Finally the most sophisticated creature of all, man, began *his* return to the sea.

Today man has the technology to enter and remain in the sea for short periods of time. He has begun interspecies communication with a sea mammal (the dolphin). He has begun to extract a living from the sea and he takes much of his recreation there. But how close is he to really living under the sea? How close is he to becoming a miner or farmer and actually *harvesting* the sea's natural resources?

Let's begin with a story—a word-of-mouth record of what must have been one of man's very first invasions of the sea. This event took place more than two thousand years ago.

2 · *The Early Submersibles*

The day had begun, and a pale crocus-yellow light washed away the last of the darkness.

A single-masted, double-ender boat seesawed gently on the low, near-shore Mediterranean waters.

Aboard the vessel were perhaps half a dozen men. All were staring at a curious egg-shaped object lying on the deck. It was approximately six feet by five feet and was made of the same kind of metal used at that time for battle armor. On its sides were four rather large oval-shaped windows of a transparent material, most probably ground and polished tortoise shell. A kind of chamber pot lid, tipped on end, made an entryway into the forbidding interior. Two stout hawser lines, attached to either end, were apparently to be used for lifting.

A man of some fifty years, with a thick flowing beard and the quiet assured manner of one who believes in what he is doing, spoke quietly to a younger man standing at his side.

The younger man was probably in his early thirties. His body had the straight, strong look of a seasoned soldier. The proud, sure carriage of his head bore the certain stamp of nobility. His eyes were remarkable, amber-green, like those of a restless tiger.

The bearded man completed his instructions, and the younger man moved toward the chamber which was to transport him under the sea! Gracefully he slipped through the small opening and signaled for a waiting man to secure the lid.

The older man took command. He nodded to two strong men, who braced their feet, took up on the hawser lines, and slowly lowered the sealed capsule into the water. The instant the water touched the object the older man inverted a small hourglass. The sand began to flow from the upper vial into the lower.

All was silent. The men lined the gunwales, peering into the transparent water for a sign of bubbles, which would indicate a leak. There were none, but they all waited impatiently.

The sand ran out; the signal was given to raise the capsule containing the first man to observe the forbidding underwater world.

The time was late spring, the year approximately 324 B.C.

The young man was Alexander the Great, king of Macedonia. The bearded one was Aristotle, his teacher.

A short time later Alexander fell ill of a mysterious fever from which he never recovered, but which had nothing to do with this incident. A fortnight after the illness, he died—before reaching his thirty-third birthday.

Was this remarkable episode in Alexander's life fact or legend? We do not know for certain. We do know that civilization owes him much. Both he and Aristotle were constantly and prodigiously curious about the sea. Aristotle possessed an inventive mind and Alexander always had a hunger for new adventures.

For the next eighteen hundred years there appeared to be a total dearth of information regarding submersibles, but a number of recorded incidents indicate that man had not completely abandoned his bent for underwater vehicles.

In 1505 Olaf Magnus, a Swedish historian, during a fact-finding journey to Oslo, discovered that King Haakon, in a recent raid on Greenland, had brought back for public display two astonishing objects. They were sealskin submarines, each with the capacity to accommodate three crewmen.

Early in the seventeenth century an eyewitness account tells an exciting story of Cossacks attacking warships at Constantinople in tremendous cowhide submarines carrying nearly forty rowers, who breathed through tubes as they annihilated the enemy.

A remarkable Dutchman named Cornelius van Drebbel built and launched a successful, functioning submarine in 1620. This unique underwater vessel was manned by twelve rowers and had a breathing tube to the surface through which lifegiving air was forced down by means of a bellows. Cornelius was able to take it down to twelve feet, and it quickly became one of the most colorful novelties of the times. Although at first the public was fearful of going for a pleasure trip in a boat that cruised underwater, the lack of accidents made them take heart, and Van Drebbel—who had emigrated to England—made regular and profitable runs from Greenwich to Westminster for ten years.

One of the earliest military underwater craft and the first submarine to carry out a tactical attack was invented by a Connecticut Yankee named David Bushnell, who in 1776 was sent to General George Washington by Governor Trumbull of Connecticut. The general was duly impressed with the persuasive young man and sent him on to Thomas Jefferson with a letter that read in part: "David Bushnell appears to be a man of great mechanical powers, fertile in invention and master of execution. He came to me recommended by Governor Trumbull and other respectable people who were converts to his plans (a most daring one indeed, submarine attack on enemy vessels). Although I wanted faith myself, I furnished him with money and other aids to carry his plan into execution."

The object that David Bushnell put together in high secrecy at Peekskill, on the Hudson River, closely resembled the barrels later used by the adventurous to transport them over Niagara Falls. As a matter of fact, this dauntless, little vessel *was* made from large barrel staves and iron. Upon completion of construction it was promptly named *The Turtle*.

The entire crew consisted of one man, who of necessity served as captain, mate, explosives officer, navigator, cabin boy, and fortunately (since *The Turtle* was to know many nerve-wracking adversities) chaplain. The one-man crew sat with his head at the level of the conning-tower ports (windows). The forward and backward motion of the vessel was accomplished by a spiral screw (a kind of propeller) in front, which was *hand-turned*. To submerge, the pilot flooded the ballast tanks, then frantically turned a vertical, helical propeller which pulled *The Turtle* down. During his idle moments he also tended the rudder, which by some miracle, more or less managed to keep the craft on course. These were just a few of the duties the operator had to accomplish while *The Turtle* was on a missionless, leisurely cruise. He didn't really begin to get busy until he began his run on target and accomplished his primary mission. Just to get a mental picture of how the operation must have looked, let's follow Ezra Bushnell, the inventor's brother (and the only man who could be found with enough energy and stamina to operate the vessel), on a practice run.

Calm and placid, Ezra sits in the pilot seat. He holds the rudder steering bar under his left armpit while his right hand turns the forward motion screw. With only the conning port above water he speeds (at a pace just under walking) toward the target. When he considers himself to be close enough, he stops turning the forward screw crank and quickly reaches back to the ballast plunger. Yanking it upward, he allows water to enter the ballast tank. At the same time his left hand cranks the vertical screw and *The Turtle* descends like a slow freight elevator. Once more Ezra's strong right hand takes over with the forward screw and the submarine plods forward until it is in striking position, below the enemy ship. Letting go of the forward screw crank, Ezra reaches straight down and pumps some of the water out of the ballast tank. Simultaneously his left hand turns the vertical screw and he ascends with some force. The force is necessary because he must impale his auger (located above the conning hatch) into the wood bottom of the enemy hull. Connected to the auger is a rope to which is attached a casement containing one hundred and thirty pounds of gunpowder. (During the on-course trip this gunpowder casement is rigged to *The Turtle*'s stern.) Suddenly there is a jolt; Ezra drops everything else and concentrates on screwing the auger deeper into the hull above him. He strains, because this is the most difficult part of the operation. His superior strength succeeds and he trips a switch that does two things: It detaches the explosive armament from *The Turtle* (if it doesn't release, he remains screwed into the enemy hull and blows up with the enemy) and it activates a clock set to detonate the gunpowder in exactly (more or less) a half hour.

Right! Auger firmly screwed in. Release switch worked (thank God). Let's get out of here fast, Ezra. Turn the vertical screw, let water into the ballast tank, a little speed or you'll land in Davy Jones's locker.

He is O.K.! He's far enough below the enemy ship now, and he gives a mighty tug on the rudder bar which slews him around 180 degrees. His right hand seems to go berserk as it turns the forward screw, scudding *The Turtle* at the breakneck speed of just under two knots. Some time later, when he knows that he is clear of the enemy, he lets go of the forward screw crank and furiously pumps the water out of the ballast tank. His left hand seeks out the vertical screw crank and his mighty juggernaut bobs to the surface. Just in the nick of time, too, because the interior of his tiny vessel is fresh out of air for him to breathe.

When one considers what David Bushnell had to work with in the way of known empirical techniques, one must admit that his little and often-ridiculed sub was a brilliant conception.

General Washington, who had little or no faith in *The Turtle,* was in a desperate position and was willing to try it. He had no navy at all and England had the largest, most powerful fleet of warships in the world—most of which seemed to be in New York Bay.

Just two days before the signing of the Declaration of Independence, General William Howe had brought an army of ten thousand professional, well-seasoned soldiers by sea to Staten Island. Washington had already marched his nonprofessional but determined army from New England to Long Island. News of the Declaration of Independence had not reached New York. Two days later, to show the colonists the futility of a fight, General Howe's brother, Admiral Richard Howe, haughtily sailed his fleet through the Narrows. The fleet was of heroic proportions, consisting of 1,000 squaresail ships of the line and 150 transports fresh from England. Admiral Howe brought in his own 64-gun flagship HMS *Eagle*. He also brought word that if the colonists wanted war he was to aid his brother in providing just that.

General Washington saw an opportunity to divert and annoy the English with his new secret weapon, *The Turtle*. He sent word to Bushnell to attack as soon as feasible.

David Bushnell decided on a coup to end all coups. He would send *The Turtle* to attack HMS *Eagle*.

Ezra Bushnell fell ill.

David refused to be stopped by his brother's somewhat convenient illness. He was given permission to appeal to General Samuel Parsons' troops for a volunteer who would be trained to replace him. Three men came forward and were trained in submarine piloting in Gravesend Bay, practically in the shadow of the enemy's fantail. A sergeant from Lyme, Connecticut, named Ezra Lee graduated *summa cum laude*, and was chosen as the next official operator of the first wartime submarine.

It was then the end of August, and the Howe brothers launched an assault on New York, ferrying twenty thousand troops from Staten Island to Gravesend Beach, Long Island. *The Turtle*, of course, just happened to be right in the path of embarkation.

The determined Bushnell performed a frenetic escape. He and his men hauled the boat out of the bay into a large wagon, pulling her overland to Whitestone, just ahead of the oncoming, red-coated British. With a quick splash, the small sub went into Long Island Sound under the shroud of night and then was towed off to New Rochelle.

At the same time General Howe pushed Washington's army into Columbia Heights, while he himself settled his army for a siege in nearby Flatbush. The wily Washington, however, furled his tents and in the dead of night quietly stole off into Manhattan. Admiral Howe then moved his fleet to the north of Governor's Island on the East River.

Bushnell's small team of men observed the fleet movement from a hidden vantage point. They devised a masterly plan to tow *The Turtle* past Hell's Gate, then let her drift downriver right into the British anchorage. Even at night the *Eagle* could be recognized in the fleet because she had the largest silhouette. Bushnell figured out everything—even the return trip, when the tide would be on the ebb, aiding the victorious sub to get away safely.

Ezra Lee prepared to make the historic trip—or to meet his Maker.

On a calm, perfect evening, the submarine was lashed to three canoes and the silent run began at Hell's Gate. The invincible armada slid past the submerged rocks and then past Blackwell's Island, slipped between the sleeping farms of Manhattan and the Newtown marshes . . . then back-paddled and halted, for there, just ahead, was a tall, menacing forest of British masts, the glint of brass cannon, almost bright in the moonlight . . . and all waiting for Ezra Lee, commander of the mighty *Turtle*. The canoes cast loose and watched for a moment while Lee flooded his tanks and submerged to the level of the conning-tower window. Then they turned and dipped their paddles in the direction of New York.

Completely alone now, Ezra Lee found that the ebbing current was somewhat stronger than he had expected and it was veering him away from HMS *Eagle*. He pulled powerfully on the tiller, turned the forward propeller with leviathan strength, and gradually he compensated for the frustrating tide. He struggled with the tide for two and a half hours, making little progress in the direction of his target. Then the tide slackened; he moved with some ease, and he suddenly became aware of what was happening around him. He heard dimly many ships' bells and saw the ships' lanterns begin to light up the darkness . . . but now the *Eagle* loomed up directly in front of him. Just above he could see the glow of lanterns coming from the captain's cabin. He began quickly the complicated, energetic process of submerging. Down he went, down, down. The *Eagle* drew almost eleven feet of water, an incredible depth for *The Turtle*. The timing was perfect. The tide was slack now, not moving at all, and *The Turtle* easily moved into place so that the auger, which was attached to the explosives, could be screwed to the hull. Lee pushed the screw into the hull and threw his great strength into driving it deep, but something went wrong. The auger refused to bite; instead it dragged his craft around. He tried again and heard a dull clang, and only then did he realize that the bottom of the *Eagle* was sheathed in copper, doubtless to protect her against teredo worms. Perspiring and frustrated but determined, Lee bounced *The Turtle* off the bottom several times in an attempt to penetrate the metal. He was making a frightful racket and he snatched a lucid moment from his rage to wonder whether or not he was giving the captain nightmares.

Suddenly he felt himself sliding toward the surface, and he realized that he had lost his trim. He popped up in the middle of the fleet and saw with horror that the ships were bathed in the first golden glow of dawn. He took no time to admire the sight, but submerged immediately and headed out of New York harbor.

Ezra Lee had many other adventures aboard *The Turtle* but, alas, was destined never to sink a ship of the Royal Navy.

The Turtle, however, was a proud parent, and she brought forth a progeny that developed into a force to be reckoned with in later wars.

In 1800, Robert Fulton, a Scots-Irish farm boy who had been born near Philadelphia, invented a 21-foot, torpedo-shaped, wooden-hulled submarine that he called the *Nautilus*. In 1806 he built his famous steamboat, the *Clermont*. His was not the first steamboat, however; at least seven practical steamboats preceded it. His great passion was always underseas craft, and in 1815 he built a submarine called the *Mute*. She was 80 feet long, had a 22-foot beam, and carried a crew of one hundred men. The crew worked the hand propeller which operated the craft. Fulton was the first user of a four-bladed screw instead of the spiral screw used by Bushnell. Unfortunately the inventor died before the *Mute* could be launched, and because of narrow-mindedness and lack of visionary men, passed into oblivion in the naval arsenal.

In 1820 a Hampshire man and ex-smuggler, Captain Johnson, built a 100-foot submarine for the purpose of liberating Napoleon from St. Helena and bringing him to the United States. He did not succeed.

In 1849 a twenty-eight-year-old corporal in the Bavarian Light Horse Artillery, named Wilhelm Bauer, built a 25-foot submarine at Kiel, Germany. She was made of sheet iron and built into the graceful form of a porpoise. She was named the *Brandtaucher* (Sea-Diver) and was the second submarine to attack in war. In 1848 the Danish navy raided and blockaded the German coast. In

December 1850 Corporal Bauer took the *Brandtaucher* against the Danish blockaders, who broke and scattered at the mere sight of her.

In approximately 1875 a man named Oliver Halsted sold a submarine to the U.S. Navy called the *Intelligent Whale*. Somewhat less intelligent than its namesake, the awkward, impractical craft managed to drown thirty-nine men during tests. Mr. Halsted's own career was shortened during the performance of his extracurricular activities. He was shot by his mistress's lover.

During the year 1869 Jules Verne invented the fictional submarine called the *Nautilus*. The ship that directly inspired Verne was the *Plongeur*, built in 1863 by Simeon Borgeois and Charles Marie Brun. She was a 435-ton craft, 140 feet long, running on a compressed air machine. The *Plongeur* proved to be a vital experiment; it led to the construction of the *Gymnote*, launched at Toulon in 1886. She was planned by the first competent, professional submarine designer, Charles-Henri Laurent Dupuy de Lome. The *Gymnote* was a 30-ton streamlined boat, 56 feet long. She ran on 54-horsepower electric motors at seven knots on the surface and four knots underwater. She had a five-man crew, a periscope, and electric lights.

John Phillip Holland, a thin, Irish ex-teacher, began building submarines in New Jersey in 1878. Altogether he built some nine submarines, won the American competition for submarine design, sold the U. S. Navy its first submarine, and with Isaac Rice established the Electric Boat Company. The same firm launched the atomic *Nautilus* in 1955.

Also in the late 1880's in New Jersey, a youngster named Simon Lake was reading every book about submarines he could get his hands on, including *Twenty Thousand Leagues under the Sea*. In 1894 he completed work on an underseas vessel called the *Argonaut Junior*. It was a 14-foot, flatiron-shaped vessel made of two layers of yellow pine with waterproof canvas between them. The propeller was hand-turned and the craft boasted a set of wheels to run along the ocean floor. For the air supply Lake had purchased a large compressed-air tank from a bankrupt soda fountain. His greatest idea, and one that really worked, was an airlock in the bottom of the boat, through which his crew could duck out in bucket helmets. Simon Lake took the *Argonaut Junior* to New York City in 1896, anchored it in New York Bay, and bewitched the entire city. His next boat was the 36-foot iron *Argonaut*, with 7-foot cast iron wheels and a gasoline engine. Although Lake had to keep her to periscope depth, to keep the air tubes that fed the gas engines and exhausted fumes out of the water, he did manage to keep her down for ten hours.

Lake was obsessed with submarines, but he was a far-seeing man, and as early as 1898 he advocated huge cargo and passenger submarines to link the continents of the Northern Hemisphere by the shortest sea route—under the polar ice!

He managed to sell the Russians five submarines but never had one accepted by his own United States. Lake moved to Berlin, where his designs were politely stolen.

On April 11, 1900, the U.S. Navy accepted what was to become the prototype of the modern submarine. It was John Holland's ninth design and incorporated nearly all the control, operating, and design features of later submarines. Called the *Holland*, she was 53 feet long, 10 feet 7 inches in diameter, and was able to cruise fifty miles submerged at a speed of six knots.

By the beginning of World War I France had 77 submarines; Great Britain, 55; the United States, 38; and Germany, 28. Although Germany had the least number of boats she used them most efficiently and by 1916 had sunk a total of 250 British ships. She immediately launched a program to increase her fleet.

From then on practically all submarine development was slanted toward military use. Early in 1942, during World War II, German U-boats were at their most effective. Twenty-five ships were sunk near the mouth of the Mississippi River and an equal number were destroyed off Florida. Shipping suffered depleting losses in the Caribbean and off the North Atlantic coast of the United States. Mines were laid by the U-boats with the result that New York harbor was closed to all traffic for three days in November and Chesapeake Bay for longer periods in June and September.

The average U-boat sank twenty ships during her fighting lifetime.

At the end of World War II the German submarine was in a state of transition. The Germans were already in the process of installing snorkel breathing tubes in their vessels. These tubes enabled submarines to cruise submerged, which permitted charging of their batteries while only the small surface of the tube was exposed. This began a search for the design and development of the "true" submarine—one that could stay submerged for long periods of time without surfacing.

It was in this respect that the Americans achieved tremendous success and surged ahead as world leaders in submarine design and building.

Admiral Hyman G. Rickover of our own Navy had to fight hard against naval and political cliques, was twice passed over for promotion by the Navy Selection Board, and very nearly forced to retire in the midst of construction of what is now recognized as one of this century's greatest achievements—the atomic submarine, *Nautilus*. His tough, vituperative battle went on for four years before the keel plate was laid at Groton, Connecticut. Finally, on January 17, 1955, the signal that marked the first use of atomic energy for propulsion was flashed from the quartermaster of the *Nautilus* to the quartermaster aboard the USS *Skylark*: "Underway 1100 hours on nuclear power."

In July 1968 Admiral Rickover once again made news with the ominous prediction that the Soviet Union has in recent years made underseas vessels the strongest arm of its navy. The Admiral claims this is especially true in the area of nuclear attack submarines.

Admiral Rickover feels that we have lagged so far behind Russia that the resulting situation poses a direct threat to our national defense. "I think it is probably too late even now," he told the Preparedness Investigation Subcommittee of the Senate Armed Forces Committee, "but I think if we work hard we can try to catch up."

England, too, is concerned about Russia's fleet of 400 submarines, which is more than double the combined strength of all the Western allies. Sir Barnes Wallis heads a project to increase his country's fleet, especially with a revolutionary new type of nuclear-powered deep-diving cargo-carrying vessel. This sub will be capable of carrying cargo and crew great distances at a depth of 3,000 feet and a speed of forty knots, allowing men and supplies to be carried thousands of miles with scant chance of detection.

3 · *The New Submersibles*

The increasing sophistication of submarines in just the last decade has been incredible. Submersibles, from the military type to search and recovery types, have developed at an astonishing rate. We have already proved our ability to search for and recover objects lying on the ocean bottom at depths of 3,000 feet. We can send men and equipment under the polar ice cap, or they can cruise effortlessly over a vast abyss 10,000 feet deep.

But until very recently there was one weak chink in the armor: We had no capacity to search and rescue beyond a depth of 400 feet!

The General Dynamics Corporation, through its Electric Boat Division, has designed and built the colossuses of our times—the Fleet Ballistic Missile Weapons System, better known, after the name of its deadly missile, as the *Polaris* fleet. Among the fleet are mammoth submarines such as the USS *George Washington*, 380 feet long with a beam of 33 feet and a displacement of 5,900 tons; the *Ethan Allen*, 410 feet long with a displacement of 6,900 tons; and the *Lafayette*, with a length of 425 feet and a displacement of 7,000 tons.

All three classes are driven by steam turbines powered by a water-cooled nuclear reactor, which gives them tremendous range and speed. Already their exploits have become legend. They have stayed submerged for months. They have crossed under the Arctic ice cap. These ships also provide a strong arm in both defensive and offensive weaponry because they can launch a mission destructing missile without having to surface. They possess the much-needed ability to strike fast with withering force while maintaining maximum security as far as detection is concerned. They are away before the enemy knows he has been hit, and so far no weaponry exists which can "seek out" these fast-moving targets and destroy them. Certainly this is the ultimate in man's genius for design and engineering.

In 1963 the submarine *Thresher* went down into the abyss with all aboard. Although frantic search efforts were made, it took a long time to find even her wreckage.

Our nation was struck with the numbing fact that we had no capabilities to search and rescue beyond a depth of 400 feet! We were sending men and equipment on military and other maneuvers in depths greatly in excess of our ability to rescue them. In May 1968 the atomic submarine *Scorpion*, with a crew of nearly one hundred, disappeared without a trace. After five months of search, part of the hull of the *Scorpion* was located in more than 10,000 feet of water 400 miles southwest of the Azores. The wreckage was discovered through the use of deep-water cameras operated remotely by the U.S. Navy oceanographic research (surface) ship, the *Mitzar*.

The *Mitzar*, commanded by James G. Hobbs, did a masterly job of marine detective work by retracing the course of the ill-fated *Scorpion* from the time she gave her last radio report until the approximate time of her disappearance.

Aware of the need of search and rescue submarines, the Navy and private industry diligently developed equipment until they could point with some pride to quick accomplishments. In just three years, the new techniques were put into action.

In January 1966 two U.S. aircraft met in a violent and terminal collision over Spain, spilling a number of unarmed nuclear weapons over the countryside, and one into the deep water near Palomares. This called for a feverish search utilizing both civilian and Navy experts.

Ray Pitts, vice-president of Ocean Systems, served as chief civilian advisor to the Navy. Jon Lindbergh, son of aviation pioneer Charles Lindbergh, was Mr. Pitts's assistant.

A number of deep-sea submersibles were used for the search, but it was the fabulous *Alvin*, owned and operated by the Office of Naval Research, Woods Hole Oceanographic Institution, which located the missing bomb in 2,850 feet of water.

The subsequent recovery of this bomb in such deep water pointed out that we had acquired some limited capacity in three years. What should have been a minor task, however, required three months, thousands of people, several dozen ships and aircraft, and more millions of dollars than we care to think about.

Since then the Navy and private industry have carried out a number of missions in deep water, successfully recovering downed aircraft, torpedoes, and other valuable equipment. Most valuable thing of all, they have gained a considerable amount of practical experience.

Except for the *Scorpion*, to date the deepest recovery has been from approximately 3,000 feet. About eighty-two percent of the world's waters are in excess of that depth, which we have only just begun to plumb.

In 1968 the Navy began work on the development of a small manned submersible with rescue and recovery equipment capable of operating at depths of 20,000 feet.

Most important of all, the Navy is developing a new system for rescuing personnel from sunken submarines. This will replace the rather ancient McCann bell, which was developed about thirty-seven years ago. The main element of this new system is a 50-foot, multichambered, 30-ton rescue vehicle capable of removing twenty-four people from a sunken and disabled submarine.

The small research submarine is developing apace. Electric Boat's first product was called *Star I*, and it participated in Project Sealab, the Navy's underwater living experiment off the coast of Bermuda in 1964. *Star I* was a one-man test submarine designed for coastal operation at 200-foot depths. She was 10 feet 2 inches long, 4 feet in diameter, made of ⅜-inch steel, had a four-hour endurance, and a top speed of one knot.

Star II, built for the University of Pennsylvania in 1964, was destined for a much more romantic and useful life. She was subsequently christened *Asherah*, after the Phoenician goddess of the deep, and in cooperation with the National Geographic Society has been used extensively for archaeological exploration off the Turkish coast.

The adventurous *Asherah* has a maximum operating depth of 600 feet, a speed range of one to four knots, houses one operator and one observer, is 17 feet long with a hull diameter of 5 feet, is constructed of ⅝-inch steel, and has an endurance of ten hours underwater under normal operating conditions.

Star III, looking for all the world like a sleek Mako shark, although still somewhat experimental, is a multipurpose vessel. Operating at 2,000-foot depths and with speeds up to six knots, she is very well suited to search and recovery work as well as the essential task of underwater mapping. Visibility is provided by viewing ports in both the trucked hatch and the main hull. The craft is also equipped with cameras, exterior lights, an exterior manipulator arm, and other instrumentation, and is capable of a submerged endurance of twelve hours. I was aboard the *Star III*, and although much impressed with her functional design, I felt the size and positioning of her viewports could be improved.

Union Carbide Company and General Precision Equipment Corporation, through their subsidiary, Ocean Systems, Inc., is another pacesetter in our search for total ocean capability. This company's contributions to underseas development are many and varied, and thus will be mentioned a number of times throughout this book, as we discuss other aspects of the "man in the sea" program.

Unquestionably one of the most imaginative deep submersibles built and tested in recent times is the *Deep Diver*, conceived and designed by Edwin Link, Chief Marine Consultant to Ocean Systems, and John Perry of Perry Submarine Builders, Inc.

The *Deep Diver* is a 22-foot long, 5-foot beam, 8-ton, dual-compartmented, completely unique submarine designed for an operating depth of 2,000 feet. It has a submerged maximum cruising speed of three knots for thirty minutes, two knots for six hours, or half a knot for twelve hours.

The most unique advantage of this submarine is its "lock-out" chamber, a separate compartment which can be pressurized to match the ambient water pressure, allowing a hatch to be opened. The diver can then enter the water and perform a nearby task while the submarine parks a few feet away. After the job is completed, the diver reenters the lock-out chamber and closes the hatch. At this point he can start his decompression or be transported to another work location for another quick job.

Working with the Smithsonian Institution, the *Deep Diver* performed a number of tasks in the Bahamas with flawless accuracy. On the first project, a technician accompanied by an Ocean Systems diver set out three light stations at a depth of 150 feet. Using the submarine as a base of operations, the men made periodic visits to each station to observe fish reactions to the different light sources. The experiment took place in a little more than twenty-four hours.

The second project took two divers to various locations on the continental shelf at depths of 50, 100, and 150 feet. At each location the men looked out of the boat and conducted a survey of marine life over a period of twenty-four hours.

In both experiments the divers reentered the chamber to decompress for approximately thirty-six hours.

Several other large corporations in the vast complex of private industry have turned their skills to the design and manufacture of deep-water submarines. Lockheed Aircraft Corporation, for example, has the remarkable *Deep Quest*, with an operating depth of 6,000 feet. Westinghouse has the *Deepstar 2000* and the *Deepstar 4000* (the numbers indicate operating depth). General Motors has the *Dowb* with a design depth of 6,500 feet, and Reynolds International has the *Aluminaut*.

It is the Grumman Aircraft Engineering Corporation, however, that, in combination with Dr. Jacques Piccard, has come up with an exciting new concept in submersibles and an adventurous first mission that will make headlines!

The vessel is called the PX-15. It is 48 feet long with a beam of 18 feet 6 inches. It has an operational depth of 2,000 feet, a collapse depth of 4,000 feet, a maximum submerged speed in excess of four knots, and it will support six men for four weeks, plus two weeks' emergency reserve. The hull has twenty-nine viewports to provide all-around visibility.

The first mission of PX-15 should tempt that adventurer-author of yesteryear, Richard Haliburton, to return from beyond the veil. Speaking from a personal point of view, I would give up a first trip to the moon to make this journey.

Dr. Jacques Piccard, accompanied by five associate scientists, will make a four- to six-week voyage from Florida to Nova Scotia submerged in the most affluent of all underwater rivers, the Gulf Stream—a historic odyssey of 1,500 miles.

What wonders they will see! Sharks, larger than man has ever seen, will death-stalk schools of fish; giant manta rays will block out the light from the sun as they hover over the small craft; vast schools of purple and orange dolphin will swim by like guests attending Mardi Gras; porpoises will pause at the viewing ports to look in on the observers who are watching them; mackerel, king fish, barracuda, will swim by in streaks of purple and silver; tuna will soar overhead like a squadron of blimps. A never-ending parade of color and motion, a teeming activity of sea creatures.

Then the penultimate—the sea life that has never been seen by man before! It will most certainly be observed by Dr. Piccard and his crew, for they will hover at varying depths from 300 to 2,000 feet. Utilizing the twenty-nine ports and high-intensity light they will be able to make constant observations.

Underwater vehicles are also being developed for civilian, recreational, and family use.

Pacific Submarines of Honolulu now offers excursions 1,200 feet under the sea off the island of Oahu for $2,500 per trip. At the moment, this type of trip is for the "man who has everything" but it does indicate a trend.

As the underwater region becomes less hostile and the world population spills over its limits of land masses, we must more and more take up some kind of habitat on and under the sea.

At a not-too-distant date we will have a complete line of underwater vehicles such as family sedans, station wagons, jeeps, and buses for transportation from one place to another.

Even today one can purchase a submarine built for two which is perfect for a quick spin to the nearest coral reef.

With all this activity it becomes obvious that measures must be adopted for control of these submersibles, just as we have done with boats, airplanes, and cars. The Navy has already instituted a certification process. All subs must live up to equipment, safety, and performance standards, as must their pilots.

We have presented a comprehensive picture of the development of submersibles and their role in the present and future of oceanography. We have with deliberate intent left out one type of submersible because it deserves a chapter to itself. Its role in oceanography is a formidable one. It is a mammal known as the *Tursiops truncatus* (bottle-nose dolphin), and the next chapter will be dedicated to a discussion of his activities.

4 · My Friend "Flippa"

The people bobbed up and down like typewriter keys. Multicolored monorails slid by overhead on their sightseeing missions. Cameras clicked like an IBM computer gone mad. Voices—of men, women, children, low, high and screeching—combined to make a carnival symphony. A sudden gust of laughter swept like a gale across the area. A public address system blared with abrupt asperity, "The next show starring the famous 'Flipper' is about to begin in the main tank."

The throngs disappeared like ants into the myriad passageways of nearby buildings, and I found myself alone on the sun-hot paths of the Miami Seaquarium.

I had made arrangements to meet two people—Mr. Ric O'Feldman, trainer, and Mr. Flipper, *Tursiops truncatus*. I use the word "people" when referring to the *Tursiops truncatus*, or bottle-nose dolphin, not tongue-in-cheek, but in all seriousness, for this truly remarkable creature has a brain larger than man's. In most cases it also has better, more peace-loving instincts.

Ric O'Feldman's first friendly act was to rescue me from the not too friendly embrace of a huge canine, as I knocked at the door of his office. It was a wolf-sized German shepherd.

"Sa'll right, he won't bite. Come on in." Mr. O'Feldman is a tall, mild, pleasant-looking young man. I decided he looked completely trustworthy, so I accepted his invitation and entered his somewhat rustic office.

I sat down on a large piece of furniture, which a very long time ago had probably been a couch, and pen in hand assumed my best reportorial manner.

It was at this point that I was attacked by another of Mr. O'Feldman's "friends." This time it was an opossum—an almost tame but quite large marsupial.

"Sa'll right," said Mr. O'Feldman, "she won't bite."

I really wasn't too worried about that. The most disconcerting thing was that "she" kept attempting to steal my pen. When I tried pulling it out of her mouth, she warned me with a malicious look of her beady little eyes and a very nasty snarl.

Finally, I had had enough. I yanked my pen out of her toothy grasp, snarled back at her, and swept her off my lap onto the floor. She gave me one last injured look and slunk away.

Ric O'Feldman is one of the dolphin and sea lion trainers employed by the master animal trainer of them all, Ivan Tors. Although Ric has only had a few years' experience, he has a great deal of natural talent for this kind of work, plus a complete and touching devotion for the dolphin.

As a matter of fact, Ric has started an organization called "Protect the Dolphins Project" whose address is P. O. Box 482, Coconut Grove Station, Miami, Florida. Practically all his spare money goes into keeping this project alive. After writing countless letters, he has acquired an impressive list of eighty personalities who have agreed to help "save the dolphins."

"It is a serious problem really, Jim," he said. "They are not protected by law. Anyone can go out in a boat and capture them or slaughter them. Too often the ignorant, amateur so-called sportsman does kill them."

"Have you written to your congressman about this and given him the information?" I asked, as concerned as he about the situation.

"I have written to as many congressmen as I thought would respond."

"Any replies?"

Ric pawed through his file and produced a few letters, handing them over to me.

I read through them quickly. Except for one serious letter, they were written in the placating tone of the politician who has no genuine understanding or interest.

"You still have a problem, buddy," I said sympathetically.

O'Feldman said nothing for a moment but sat in a brooding silence. Then he stirred and said, "Let's go out and check on the dolphins."

We threaded our way through the crowds as we headed for the dolphin tanks. They were located in an isolated part of the park where only working personnel and invited guests are allowed. Fortunately for us, en route we ran into Don Renn, whom I had known for years and whom I consider to be one of the best underwater photographers in the United States. Don agreed to accompany us and take some pictures as I met Flipper for the first time.

"Flipper" turned out to be Suzy. As a matter of fact, there were two females in the training tank. They greeted Ric with a flurry of enthusiasm, whereas they viewed me with some definite, albeit good-natured, suspicion.

"They really are remarkable, Jim," said their trainer. "They can learn the most complicated tricks in seconds. As a matter of fact, very often they will teach *you* a trick. Out of the clear blue they will perform something they have never done before, then wait to get your reaction. If you like it and feed them, they keep the new trick in their repertoire."

"Ric, as you know, I've spent much of my life on or under the water. I have swum with dolphins many times out in open water, but there is something I have always wondered. Their behavior always seems to be the same. Is there any variation in their temperament?"

"Absolutely." Ric smiled and looked fondly at the frolicking mammals. "Although they are basically loving and good-natured, I have seen them annoyed, impatient, even jealous."

"I can do without your dog and 'possum, Ric, but I must say I like your girls. When may I start to work with them?"

"Well, they have already become used to your presence, so you might as well start right in. The first thing you do is take this pail of fish and feed them."

Within the hour Suzy, Sally, and I were working together like a vaudeville team. I learned to feed them without worrying about their extremely sharp teeth, shook hands with them, put my arms around them, and had them retrieve various objects from the bottom of the pool.

After the first flush of success had begun to pale, Ric and I decided to attempt the ultimate.

Ivan Tors's Florida studio was shooting a feature film entitled *Hello Down There*, which is the story of the first family to live in an experimental underwater house. There is a scene in which the father of the family finds it necessary to accomplish some repair work on the outside of the house. In the middle of the task he discovers that he requires a certain tool. The family's pet dolphin is playing nearby, and he is sent back into the house to fetch the missing tool.

We simulated the situation in the small training tank at the Miami Seaquarium. Within two minutes Suzy was performing the trick with one hundred percent accuracy.

The dolphin's "finding sonar" is even more accurate than its ability to memorize. We threw pennies to the bottom of the murky, algae-thick pool, to have them picked up and returned in a flash. Dolphins possess a built-in sonar system, better than the electronics systems in any submarine. They send out a stream of "clicks" which become, in effect, a sonar "signal." This signal finds the target and bounces back to the direction-finding "receiver" of this remarkable mammal, and he speeds directly to the object.

Toward the end of that afternoon, I noticed that whenever Ric and I paused for an exchange of thoughts and conversation, Sally would splash some water at us with her flipper. The first few times I thought that it had been an accident. A natural vocational hazard, so to speak. If one played with dolphins one must expect to get drenched. Then I noticed that Suzy never, at any time, splashed us. The intermittent deluge was precipitated entirely by Sally. During one of our pauses, I watched for it, and actually caught Sally in the act. She poked her head out of the water, her eyes agleam, waiting for our next move. Ric and I ignored her, but I continued to watch out of the corner of my eye. Sally's expression appeared to change while she waited impatiently for us to continue to play. Nothing happened. She moved her head back and forth, sloshing the water. Still no excitement. Then I saw her do it. She pushed her body farther out of the water, and scooping up a half gallon or so of water, flung it with deadly aim directly at us.

Ric grinned. "Guess I'm not such a good trainer after all. I really should do something about that. Still it's her only bad habit. I thought I'd let her keep it."

I decided that I wanted to learn more about the remarkable *Tursiops truncatus*, and I knew a spot where I should be able to observe them in open water.

There is a cove between Greyhound Key and Marathon in the Florida Keys, on the ocean side, where I had seen more dolphin activity than I had ever seen anywhere else. I had discovered the place a few years before while filming an underwater movie there.

I looked up George Greer, a Florida diving buddy who would drop everything at the suggestion of a diving trip. We gathered up diving gear, rented a small boat with an outboard motor, rented a small tent, bought food supplies, and set out for the cove.

The cove was marked by a tree that had become partially up-rooted and stretched its length over the water. We found it easily, set up camp, then took the boat out about a thousand feet offshore and anchored there.

About two o'clock that afternoon George and I spotted a small school of dolphins gamboling in the water a thousand feet farther out. We started the outboard and headed in their direction. As we neared the spot George cut the engine and we drifted in. Quickly donning mask, fins, and snorkel, I slipped into the water. Reluctantly, George agreed to skipper the boat.

The water was clear and I saw a school of six or seven just off to my right. They were feeding on what appeared to be a large school of mullet as I approached. Since my approach was from the surface I presumed that my presence had gone undetected. I kept my fins undulating slowly, propelling my body forward until I was directly overhead. Something happened then to make me realize that I had not glided in unobserved, but had indeed been under constant surveillance.

One of the feeding dolphins detached himself from the others and headed directly up at me with the speed of an express train. His mouth was wide open as he charged.

At the last moment he swerved, leaped clear out of the water, and entered again on the other side of me. He rejoined the others, emitted a series of clicking sounds, and then the entire brood took off like rabbits.

On the surface George Greer was eloquent about the sight of the dolphin jumping clear over my snorkeling body. He asked what I intended to do for an encore.

The next afternoon we anchored the boat in the same location, but this time we were equipped with a pail of fish chunks, which we stacked in our king-size ice chest.

The dolphins appeared on schedule and we began to tempt them with the fish. I held a choice piece of bait just above the surface of the water, but none of the mammals approached it, and finally I let it slip into the water. A body streaked by and snatched the piece before it hit the bottom.

George and I both went into the water this time, but equipped with tanks so we could watch the circus from a better vantage point. We let ourselves settle to the bottom, in about thirty feet of water, and watched fascinated.

Almost as though they were performing for an audience, the dolphins cavorted like a trained team of gymnasts. They shot to the surface, disappearing completely, then shattered the mirror of the water as they reentered. They twisted, somersaulted, and swam on their backs in a frenzy of continuous activity. At one point I crept forward, pulled out a small purple seafan, and deposited it on the bottom.

One of the troop scooped up the fan and carried it to within a few feet of the surface, at which point he let it go and watched as it drifted down toward the bottom. Eyeing it carefully, he waited until it was close to the bottom, then darted down and recovered it, wearing it on his head like the lace cap of a kindly grandmother.

It was on the third day, same place, same time, when a few of the dolphins—three who were daring enough to try it—allowed themselves to be fed from the boat.

Once again George and I got into the water and took our seats at the bottom. This time, however, I brought an old canvas fishing hat. The getting-acquainted time was much shorter today, and when I held out the hat one of the dolphins actually approached and took it from my hand. The frenetic play began immediately. The hat was fair game for all. The dolphins picked it up, let it go, caught it again, blew bubbles into it, passed it back and forth like a limp baton.

Something else happened which mystified both George and me. At the height of their play, the dolphins began to mouth each other's bodies gently with their sharp teeth. With the mouth wide open, one would bite softly along the entire length of the other's body. No injuries occurred, and the recipient of the mouthings seemed rather to enjoy it.

The thought struck me that this was some kind of sex play, and I later discovered by reading some of Dr. John Cunningham Lilly's books on the behavior of dolphins that this was indeed the case.

As human beings we can "see" any image which the eye encompasses in an air atmosphere. We see shapes, colors, and designs in great detail. We can do the same thing underwater providing we are in clear water and we are wearing a glass mask that establishes an air gap between the water and our eyeballs. Place us in any type of "fog" atmosphere and we are virtually blind.

Not so with the dolphins. They can see in the air atmosphere as well as in "fog" atmosphere. In murky water the dolphin merely sends out a highly developed ultrasonic beam which "sees" the object and sends back a "picture" to his range and echo recognition mechanism. To "see" the object in the opaque water, he adjusts his sonar beam up and down, back and forth, or laterally, by movements of his head. He "sizes" the object by adjusting the sonar cone, narrowing it for small objects and widening it for large ones. Furthermore, he can take a comprehensive picture of his surroundings—front, side, and back—to a distance of several hundred yards.

We know that these mammals communicate with each other with great facility, probably with a much better sense of understanding than we. It is possible that the language of the dolphin has a more definite, acute meaning than the language of man.

Their language, however, is made up of acoustic pictures, while ours is made up of visual pictures.

Dr. John Cunningham Lilly probably knows more about dolphins than any man alive, having written a number of excellent books on the subject.

It is his contention that we can develop interspecies communication with dolphins. Dr. Lilly, his staff, and the redoubtable Margaret Howe conducted a most unusual experiment a few years ago which gave credence to a seemingly incredible theory.

Man's best friend, the dog, has an almost perfect set of vocal cords for understandable speech. He does not, however, have the ability to memorize a series of sounds in sequence to form a word. Consequently, "Fido" is committed to a life of rather monotonous barks.

The dolphin, on the other hand, has a vastly superior brain power, but his vocal equipment, for human speech, leaves something to be desired. This does not, however, rule out the possibility of acceptable humanoid communication.

The dolphin "speaks," i.e., emits sounds, through his blowhole. The sounds are formed by two separate and complete sets of phonetic equipment. To produce a variation of sounds they can be used separately or in combination, and they can also intermix the two sets of sounds. In other words, the dolphin's phonation equipment is sophisticated enough to produce humanoid sounds.

Thus it is the mind of the dolphin, shrouded now in mystery, which must be understood. To do this Dr. Lilly formulated and executed a daring and imaginative experiment.

It was decided that a staff member, Miss Margaret Howe, would live in a flooded house for ten weeks with a young male dolphin named Peter. The "living" portions of the house were flooded to a depth of twenty-two inches. Here Miss Howe would carry on normal living—cooking, sleeping (in a special hammock above the water), writing notes, watching television, answering the telephone.

The area would also serve as Peter's classroom. Other parts of the house were flooded to a depth sufficient for a dolphin to maneuver and get sufficient exercise.

The experiment took place from June 19 to August 18, 1965, on St. Thomas in the Virgin Islands.

Even during the first week Peter made some progress in learning. He was able to accomplish a good pronunciation of the word "ball" and began to imitate the inflections of Miss Howe's voice as she counted, "one, two, three, four." He preferred—and he made this preference known through his rather obstreperous behavior—the play periods, during which he fetched a ball or a rag thrown by his attractive teacher.

The second week he ceased nipping at Margaret's heels, but he did begin to conduct a more thorough inspection of her physical person, especially feet, ankles, and legs. He continued to be vociferous about his desire to play, yet made some progress with "humanoids" (human sounds, as compared to "dolphinoids," the clicking or whistling sounds dolphins generally make).

The third week appeared to be a rather neutral week during which Margaret felt her first attack of "restlessness." She had begun to miss human companionship. Peter's learning neither forged ahead nor did he become the first dolphin "dropout." Some modifications were made in the building's facilities.

A good part of the fourth week was spent outside in the deeper water. Peter attempted to stay in his teacher's good graces by talking to her in humanoids. This was something of a subterfuge, Margaret discovered, because what Peter really wanted was to play. As a matter of fact, at this point Peter began to teach Margaret new games. Recording sessions went well, with Peter listening and humanoiding.

During the balance of the allotted time Peter became more and more involved with his teacher. It might be said, if one can assume what is in a dolphin's mind, that she became "the other dolphin" with whom he shared a tank. He became more and more riotous, less and less inclined to learn, until it was discovered beyond a shadow of a doubt what it was that bothered Peter. Peter had become sexually aroused.

When this happened, teaching was of necessity suspended, and Peter was allowed a visit of a day or two with two delightful female dolphins named Sissy and Pam. After this sabbatical, school was once against successfully resumed.

The experiment uncovered a number of interesting facets. Peter did have a tremendous capacity to learn, but the teaching process was a dual one. Miss Howe had a natural fear of Peter's tremendous strength. He could, if he so decided, snap her in two pieces at any time. Peter went to great lengths to teach her that he could be trusted *not* to do this. He literally taught her to conquer this fear that could only stand in the way of his learning process.

Peter did learn a number of humanoid utterances, plus, of course, innumerable "games." Margaret Howe proved that a human can live with a dolphin for an extended period of time and that the result is an undeniable communication buildup.

More utilitarian experiments with dolphins have been accomplished in other areas, the most important of which was the Navy's underwater habitat project, Sealab II.

We will discuss this in greater detail in the next chapter. But for now, suffice it to say that this important and successful project took place in cold southern California waters off La Jolla. Teams of aquanauts lived in the habitat at a depth of 205 feet for fifteen days, making daily excursion dives.

After a few early failures a male dolphin named Tuffy, responding to buzzer signals, performed seven perfect missions in which he delivered tools and mail between surface and bottom, and with the use of a line and harness, guided "lost" aquanauts back to the safety of Sealab, their temporary underwater home.

Interest in dolphins existed long before the current, more pragmatic interest. A beautiful fresco of dolphins at play was found at the fabled palace of Knossos on the island of Crete. Aristotle avowed that the voice of the dolphin in the air was like that of a human.

But in spite of dolphins' enormous potential, as we have outlined in this chapter, these marvelous creatures are in grave danger. They are being hunted and slaughtered, since there is no law to protect them. They are being used as pets in the swimming pools of the wealthy. Unfortunately, they do not last very long this way because inadequate filters, piping, pumps, and seawater mixtures are used, along with improper diets.

If any reader feels strongly about helping to save the dolphins, I would like to suggest that he get in touch with Ric O'Feldman and his Save the Dolphin Project.

5 · The U.S. Navy's Project, Sealab

The U.S. Navy embarked on its substantially successful career in underwater habitat experimentation in a series of well-planned projects called Sealab.

The first experiment with underwater housing was Sealab I, conducted from July 20 to July 31, 1964, at a depth of 193 feet, in the clear, warm, vacation waters off Bermuda. This ideal site was twenty-six miles off the Bermuda coast, adjacent to the Navy's Argus Island (an installation for Sonar research), and the Sealab housing was to occupy a small plot of flooded real estate known to the ancient mariners as the Plantagenet Bank.

Even under ideal conditions, any experiment in underwater living is an ambitious, complicated project. More than the mere positioning of an underwater laboratory, an experiment like this calls for a well-trained aquanaut team, support divers, a shore base, diving equipment, electronic equipment, and surface craft support and crews.

Original plans called for four men to live for eleven days at a depth of 200 feet and conduct specific assignments under "saturation" conditions, i.e., breathing a mixture of gases designed for the water depth in which the Sealab was located.

Being a prototype, Sealab I was experimental in design and function and understandably left a great deal to be desired. It was constructed from two minesweeping floats, welded together to form a narrow, blimp-shaped chamber 40 feet long and 10 feet in diameter. Attached to the habitat were a complex of cables and hoses for electricity, compressed air and helium, fresh water, telephone, electro-writer, atmosphere sampling, and a two-channel television monitoring system. All the cables and hoses terminated at the surface support vessel.

The support vessel was a lighter (large ship equipped to load and unload other vessels), 260 feet long and 40 feet wide, with accommodations for a crew of thirty-two plus thirty scientists and technicians.

In the interior of the habitat, the end sections were fitted to hold water ballast and breathing gases for emergency use (normal breathing gas was pumped down from the surface lighter through the connecting umbilical). A scant 24 feet of living space was provided in the middle of the habitat and unfortunately had to be crowded with bunks, lockers, laboratory equipment, environment controls, refrigerator, hot plate, food locker, shower, toilet, air-conditioning equipment, storage space for all diving gear, and of course, the four aquanauts themselves.

These cramped quarters were somewhat deteriorative to morale and understandably hampered work missions. Larger living space was given a high priority for subsequent experiments.

An additional and important part of the Sealab complex was a Submersible Decompression Chamber (SDC) which the aquanauts and inspecting personnel used as an elevator, a safety escape chamber, and (since it was pressurized to a depth of 200 feet) a decompression chamber.

The U.S. Navy does not commit men to dangers inherent in a new experiment without thorough testing. Preliminary testing was carried out at the Mine Defense Laboratory, Panama City, Florida, and in the Gulf of Mexico. Practice lowerings were accomplished at these locations plus an exhaustive testing of all umbilicals (connections between habitat and surface control) and systems.

Finally, on the warm, quiet summer morning of July 19, 1964, a crane lowered Sealab I into the clear, blue waters while the gas mixture was pumped into the chamber to avoid flooding through its open hatches. At 1:30 in the afternoon the habitat settled level on the ocean floor at a depth of 193 feet.

Proceeding according to plan, on the following morning scuba divers went over the side of the lighter to begin the long, arduous task of securing the mooring and anchoring lines.

Captain George Bond, Medical Corps, USN, and Captain Walter Mazzone, Medical Service Corps, USN, then took the SDC to a depth of 160 feet, exited the chamber, made a final inspection of the habitat, and finding all in order returned to the surface.

At 5:30 that afternoon, the four aquanauts entered Sealab I and formally took up underwater housekeeping.

For the next eleven days the men lived in the artificial atmosphere of the habitat without the need of any special breathing apparatus. They were breathing "air" composed of 80 percent helium, 16 percent nitrogen, and 4 percent oxygen. While outside the habitat accomplishing their prescribed mission they wore the Navy's standard Mark VI, semiclosed breathing apparatus. An alternate system was the Hookah-Arawak, which provided breathing gases from the habitat through an attached umbilical hose. This gave the diver an unlimited supply of breathing gas but restricted his scope of activity to the length of the 100-foot hose. The Mark VI scuba tanks allowed the diver more freedom of movement but limited his diving time to seventy minutes.

The first night's sleep was somewhat fitful, and the next morning the aquanauts anxiously donned their scuba gear, looking forward to the first trip "outside" their underwater house. There was much more room outside their cramped quarters, and the visibility of the water was excellent. Looking up they could easily see the hull of the support vessel bobbing on the gentle waves. They closely observed the surrounding flora and fauna and made brief tours around the base of the Argus Island Tower. The fish population was dense and colorful. On subsequent days they worked at a number of pre-assigned tasks. They observed sea life, placed ultrasonic beacons on the ocean floor, installed current-measuring meters, tested shark attraction systems (they failed to attract sharks, but sharks did appear at other times), and rigged lights for night photography. In some ways there is even more activity at night among marine animals—or at least a different segment of the populace inhabits the night waters.

It was during a photograph mission that the one accident occurred which marred the otherwise accident-free record of the Sealab I experiment.

Star I, which was then an experimental one-man submersible built by the Electric Boat Division of General Dynamics, cruised into the immediate area of Sealab, on a preplanned mission. The object was to test the ability of the aquanauts to assist the craft to land on a simulated hatch of a larger submarine. Aquanaut Manning, who was a veteran diver, was photographing the procedure when he suddenly began to feel lightheaded. Deciding quickly and correctly that his scuba unit was not functioning, he turned and swam immediately back to the habitat. His gas supply had malfunctioned and he was apparently breathing his own exhalation. He had reached the hatch and climbed part way up the entryway when he lost consciousness. Gunner's Mate First Class Anderson, who was on watch in the habitat, heard the clang of Manning's scuba tanks as he fell. He investigated and just managed to grab Manning's body as it began to float away. Quickly he brought the unconscious diver into Sealab, and within minutes had him breathing normally again. Lieutenant (Doctor) Thompson returned and gave Manning a thorough examination. The blood vessels in his eyes had ruptured, causing the whites to turn red. He was kept under close observation, and except for looking like something out of the *Phantom of the Opera*, suffered no further ill effects, and so was allowed to remain as an active aquanaut.

On July 31, the eleventh day after Sealab had been lowered into the sea, it was decided to cut the experiment a day short. A violent tropical storm had been reported some distance to the north, and its fringes were already beginning to disturb the surface waters.

The order was given, and in the calm beneath the surface the aquanauts completed the tasks preparatory for return to the surface.

Original plans called for the aquanauts to remain inside and ride the habitat to the surface, using it as a decompression chamber. That way they could stay in relative comfort during the long decompression hours.

The tropical storm moving fast decided differently.

The crane on Argus Island began to lift the Sealab. As the heavy habitat neared the surface, rough seas began to buffet the chamber, causing the cables to go slack and allowing it to drop, then suddenly to get caught up again. This created monumental strains on both the cables and the crane itself.

With an on-the-spot decision, came the order for the aquanauts to abandon the habitat at 81 feet and swim into the nearby SDC. Here they would have to remain while the crane lowered the Sealab back to the floor of the ocean.

The unfortunate but unavoidable situation resulted in the four men having to stand immobile and cold, without any surface supervision possible, for more than seven hours in the tiny SDC chamber before it was hauled up and placed on the deck in a horizontal position.

Then a decompression period of fifty-six hours was begun.

Medical tests of the men proved them to be in good health. Analysis of information gathered during the experiment showed up some problem areas. Better engineering was needed for raising and lowering the habitat; lower humidity of its atmosphere was required; helium speech (breathing helium creates a "Donald Duck" voice) would have to be unscrambled; swimmer navigation equipment should be developed; larger living quarters and reduction of diving gear, or larger areas for diving preparation and dressing, would be necessary.

In spite of the problems, Sealab I was a success. It proved that man could live under the sea in saturation conditions with no ill effects.

SEALAB II

In 1965, thanks to the work and cooperation of more than four hundred people, the Navy was ready to launch the Sealab II project. It was to be a much more ambitious and sophisticated design than the pilot project, Sealab I.

The specially designed 250-foot maximum depth Sealab II habitat was 57 feet long and 12 feet in diameter, capable of withstanding an internal working pressure of 125 psi (pounds per square inch). The hull had eleven viewing ports, which were each two feet in diameter, and was provided with three access openings. The bottom entry hatch, approximately four feet in diameter, was located at the extreme end of the hull and provided the normal access and exit for the divers. An emergency hatch 27 inches in diameter was located in the hull bottom near the bow, and the surface access hatch, also 27 inches in diameter, was located in the top of the hull amidships.

The entry trunk was eight feet by eight feet and extended two and one half feet below the hull bottom. This trunk provided a displacement volume to compensate for bottom pressure variations due to tidal change as well as normal internal pressure changes. It made for a drier entry area.

A shark cage, eight feet by twelve feet, housed the entry area. The cage offered protection for the diver during the difficult unseeing moments when he climbed up or down the ladder leading to the access hatch.

The living compartment was divided into four areas. The forwardmost was the sleeping quarters with bunks for ten aquanauts, a large drop-leaf table, storage lockers, and the emergency escape hatch.

Next was the galley, which contained built-in sink and cabinets, large electric plate, chill box, freezer, electrical power transformers, and the components for the air-conditioning system.

Just aft of the galley was the laboratory, which contained a built-in sink and cabinets, a 50-gallon hot water tank, a 150-gallon emergency fresh water tank, the habitat's breathing gas control panel, and the communications station.

Finally, in the aftermost compartment, was the entryway containing the access hatch, showers, and stowage space for diving gear.

The support vessel was actually a complex of four vessels; two 110- by 35-foot YC barges, spaced 22 feet apart and connected at one end by a covered rigid platform and two LCM's (landing craft material).

The main support vessel, was completely equipped to handle all phases of the Sealab II project. Included aboard were the divers' ready room, Sealab Control Center, a deck decompression chamber, a counterweight system for raising and lowering the PTC (Personnel Transfer Capsule), Sealab, and a breathing gas storage and distribution system. Important items of machinery included three AC generators with a total capacity of 460 kw., two 1,500-pound pull wenches, a high-pressure air compressor, a low-pressure air compressor, and a 100-ton lima crane restricted to a 50-ton working load because of the way it had to be mounted.

Once Sealab II was down and the aquanauts inhabiting it, it was necessary for the support vessel to limit its motion within a ten-foot circle. This was accomplished with a five-point moor; that is, the vessel was anchored at five strategic positions.

The location of Sealab II left far behind the clear, warm waters of Bermuda. It was no longer enough to know that man could exist in 200 feet of water under ideal conditions. Sealab II was going to prove that man could perform useful tasks under more severe and hostile conditions in an atmosphere which might be considered more typical of the continental shelf. This, after all, was the ultimate aim of the man-in-the-sea program—to place man in the environment of the continental shelf so he could exploit, for the benefit of mankind, the rich resources of the ocean bed.

The site finally chosen was just off the Scripps Institute of Oceanography, north of La Jolla, California.

The water was miserably cold, 40 to 52 degrees. It was also murky, with an average visibility of about twenty feet. It was near a tricky bottom created by a submarine canyon. At several points the depth increased from 190 to 300 feet in a horizontal distance of only 100 feet.

Abundant sea life surrounds an underseas cable in 200 feet of water off Cape May, New Jersey. This photograph was taken in October 1966 from *Star II*. (*General Dynamics, Electric Boat Division*)

Dr. Harold Edgerton, developer of deep-sea camera and light systems, poses with Jacques-Yves Cousteau's submarine, *Souscoupe,* on a ship off the coast of France.

Dr. Edgerton's deep-sea camera and xenon light system is lowered into deep water by Woods Hole Oceanographic Institution personnel.

Photograph taken from *Trieste I* with Edgerton camera system at a depth of 3,500 feet shows starfish and other crustaceans. Eagle ray is partially shown at top of picture. (*Dr. Robert Dill*)

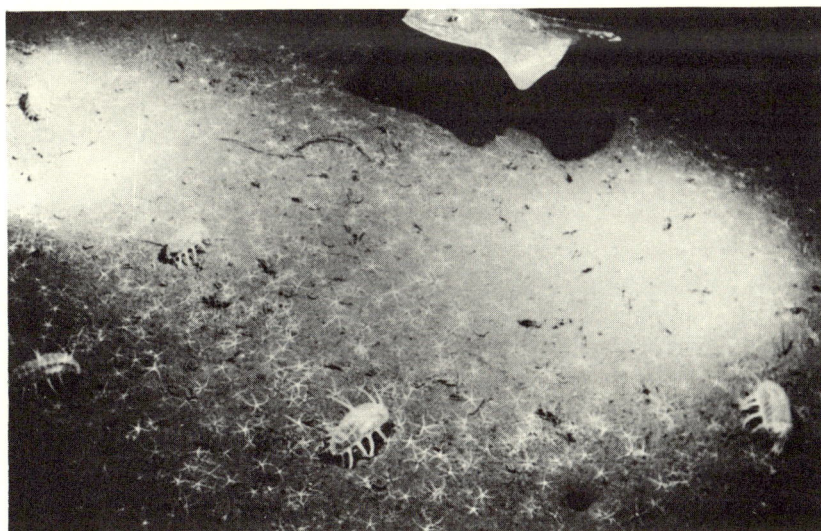

Two photographic submarines: the *Alvin* (top) and the *Trieste II* (bottom).

The author's basic tool kit includes a 16-mm. Bolex underwater movie camera, a block and tackle, a short-handle pickaxe, a chisel, and short-handle hammers.

This Neolithic sacrificial bowl, dating from 6,000 years ago, was discovered in an underwater city off the coast of Melos. In this picture the author (in rubber diving suit) poses with his find.

Trainer Ric O'Feldman shakes hands with star performer Flipper at the Miami Seaquarium. With a brain larger than a man's, the dolphin is cultivated both for its playful instincts and its ability to memorize.

Great care is taken for Flipper's comfort and safety in transit from one place to another. Here trainer O'Feldman oversees the dolphin's transportation to an on-location acting job.

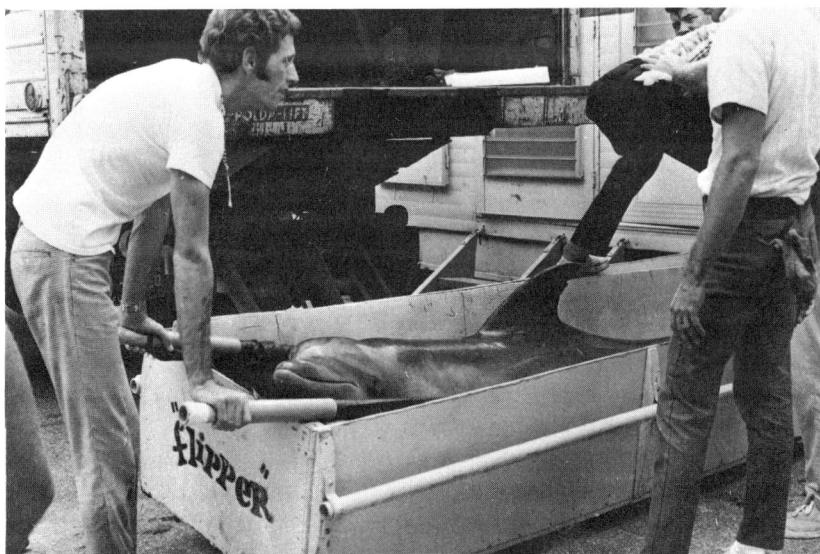

Divers on movie set for Paramount Pictures lead a tiger shark around to the camera. Film producer Ivan Tors has discovered that sharks in captivity are in a state of shock and will not attack if they are well fed.

Cinematographer Lamar Boren moves into position with a 35-mm. camera to film a scene for Ivan Tors's *Around the World Under the Sea*. Boren personally designed the pressurized camera housing that permits shooting at depths down to 200 feet.

It was a perfect site for Sealab II.

On August 26, 1965, Sealab II was placed on the bottom at a depth of 205 feet.

Plans called for three teams of ten men, with each team spending fifteen days in the habitat. Two men, Commander M. Scott Carpenter, on loan to Sealab from NASA, and Lieutenant Robert E. Sonnenburgh, Medical Corps USNR, would spend thirty days in the habitat.

The main objectives of Project Sealab II were:

1. To determine man's general ability to accomplish useful work at a depth of 205 feet in a realistic ocean environment under saturated (breathing atmosphere comparable to ambient water) diving conditions.

2. To observe whatever psychological changes might occur in a man as a result of extended diving.

3. To measure deep-water work proficiency as compared to that of shallow-water divers.

4. To determine what effects the environment's severe conditions would have on the interactions of the teams.

For psychological as well as pragmatic reasons it was essential to give the aquanauts many time-consuming tasks. This was only the beginning of a long series of projects. There were many questions still to be answered. Besides the daily tasks of standing watch and housekeeping, there were some thirty-four assignments and tests to be accomplished. A few of the major ones were plankton studies; sediment coring (drilling the bottom for sample materials); auditory range studies (measurement of the distance at which certain sounds can be heard); strength tests (individual aquanauts took push-pull strength tests to measure the difference between deep-water and shallow-water efforts); two-hand coordination tests; fish migration studies; foam-in salvage (a method of lifting wrecks through the use of air foam squirted from guns); placement and recording of weather-forecasting equipment; arithmetic tests; touch sensitivity tests; and time-lapse photography.

Besides the numerous daily chores, there was a lot of new and modified equipment to be evaluated, such as the new heated wet suits and other diving equipment. All in all, there would be few idle hours in the aquanauts' daily life.

On August 28, 1965, team number one, with Commander Scott Carpenter in charge, entered Sealab II, set up housekeeping, and began the long, arduous project. Another event happened on that first day which gave it special significance:

Aquanaut Scott Carpenter, 205 feet beneath the sea, talked via special radio link to astronaut Gordon Cooper, whose space capsule was orbiting the earth at altitudes from 106 to 217 miles. For the next forty-five days most of the work assignments were successfully carried out, studied, and evaluated.

During the entire project the men in the habitat were photographed by closed-circuit television cameras and monitored for physiological behavior evaluation on the surface support vessel.

Although this Sealab project was successful in proving that man could live and work underwater even in stressful conditions with no serious psychological damage, a number of problem areas were discovered.

There was definite evidence of psychological stress on the aquanauts during the first three to five days of living in and working outside the habitat, with a slow return to normal.

Human performance tests showed that there was degradation of work capability of between 17 and 37 percent as compared to warm, shallow-water capacity. Still, a great number of assignments were performed successfully, and the *in situ* application of work more than made up for the slight loss in proficiency.

The experimental, electrically heated wet suits were considered essentially successful, although some wire breakage did occur as well as ozone cracks (silent discharge of electricity in ordinary oxygen).

The habitat engineering features functioned extremely well except for a few minor problems: a few of the viewing ports leaked slightly; the food freezer did not maintain a sufficiently low temperature; the dehumidifiers removed moisture at less than 25 percent of their designed rate; the CO_2 scrubber was only 60 percent efficient; the entry-exit area was too small for the dressing and diving gear storage necessary for daily activity; and the noise level was too high, affecting both human comfort and communications intelligibility.

A general post-project evaluation produced the following conclusions:

Although vastly improved over the pilot project, Sealab II indicated that neither diving equipment nor habitat was ready for routine man-in-the-sea operations.

Underwater communications and ocean floor navigation are, in their present state, unacceptable for safe and successful work and living under the sea.

For a completely autonomous habitat, a reliable power unit must be developed, as well as a complete, integrated, life-support system.

A leveling technique should be developed for future habitats so they can quickly be placed level on the ocean floor.

Special viewing ports, perhaps bubble types, should be designed to provide an increased angle of observation.

Special attention should be paid to the absolute elimination of atmosphere contaminants. During the latter phase of the operation the aquanauts complained of headaches, which resulted from the presence of carbon monoxide.

A large viewing port is needed near the critical entrance area to observe divers.

The lighting in and around the shark cage should be improved.

Because of the difficulty of entry while wearing flippers, other entry techniques warrant investigation, i.e., an elevator.

A complete reevaluation of the interior of the habitat is desirable.

Photographic conditions should be improved by the installation of special photographic lights.

These and other conclusions were adapted to improve the Sealab III project.

As a result of personal observations and study, I would like to offer the following ideas as suggestions to further improve future habitat operations:

1. A large mat should be laid completely around the base and adjacent area of the habitat to prevent the disturbance of the bottom sediment, which obscures visibility.

2. A system should be developed to precipitate the suspended matter in the water around the area of the habitat. This would result in greatly improved conditions for photography and observation.

3. Numbered air stations, which can communicate to the habitat, should be installed within the scope of mission or excursion areas to aid a diver if he gets lost.

There were no accidents during Sealab II except for the case of Commander Scott Carpenter, who came in contact with one of numerous poisonous scorpion fish. His hand and arm became swollen and painful, but there were no lasting ill effects.

SEALAB III

Sealab III utilizes a modified version of the habitat used in Sealab II.

Five diving teams of nine men each, including two British navy aquanauts and several civilian scientists, were to occupy the habitat for twelve-day periods during a scheduled sixty-day experiment. Initial plans called for a two-phase operation. During the first phase the habitat and aquanauts were to operate at a depth of 450 feet. Should alert monitoring reveal no ill effects suffered by the men, the habitat would be lowered to a depth of 600 feet and the balance of the experiment carried on from there.

This plan was aborted after the successful checkout of all aquanauts to a simulated depth of 630 feet in the pressure pot located in the Navy Yard in Washington, D.C. New plans called for the direct placement of the habitat at the 630-foot bottom.

In April 1968 I went to Washington, and with Warrant Officer Robert Barth, who had participated in previous Sealab programs and will participate in Sealab III, entered the pressure pot just prior to the entry of the divers.

This diving facility was manufactured during World War II. The Navy should have a newer and completely modified experimental diving chamber, but its budget will not allow it.

The pressure pot consists of five component parts and can take a team of divers down to a simulated depth of 949 feet. If you can pretend for a moment that you are a Navy aquanaut, this is what it feels like to make a simulated deep dive.

You enter through a small round hatch into the outer lock, which is 3 feet 11 inches long and 6 feet 3½ inches in diameter. As you pass through, a hatch is slammed shut behind you and secured. From the outer lock you step through another small round hatch into the recompression chamber, or inner lock, which is 8 feet 11 inches long. These are the living and sleeping quarters, with two cots and some stacked food.

Before you have a chance to sit down, the second hatch is slammed shut with even more finality than the first.

You then crawl through a tunnel which is 3 feet in diameter and 4 feet 10¾ inches long.

You are now in what is known as the igloo, which is a metal chamber 10 feet 6¾ inches high and 9 feet 7 inches in diameter.

In the center floor of the igloo is a tank 10 feet deep, filled with murky green water.

An inert mixture of helium, nitrogen, and oxygen, carefully controlled from the outside, replaces the atmosphere you are accustomed to breathing. The wait is a long one, for all the tissues of the body must be saturated with this gas mixture before the experiment can really begin. You know you are going down when you begin to feel the pressure against your ear drums.

If there are no problems, you are brought down to the required depth for a series of tests designed to find out whether or not you are adaptable to the pressures of a deep-water environment.

At one point you put on one of the suits to be used in Sealab III. You then mount a platform and are lowered to the bottom of the murky green water, where you stay until you have completed a simulated "excursion dive." This is an important part of the test, since as an aquanaut you may be taking daily excursion dives in the deep water around Sealab III.

Finally the tedious, grueling test is over, but that does not mean the hatches will miraculously spring open and you will leave your cramped quarters to walk in the fresh air and sunshine.

You now start the long process of decompression. The inert gases have to be drained from your body tissues and bloodstream until your body and lungs can once again breathe in the lesser pressures of your normal atmosphere (14.7 pounds per square inch) without danger of your getting the bends.

So you resign yourself to another few days in the pressure pot. But it's not so bad, really. You have passed the final test. You are now a full-fledged aquanaut.

The interior of Sealab III is virtually the same as that of its predecessor. The only change is an exterior one.

Two rooms, 8 feet high and 12 feet square, have been affixed to either end of the habitat. The shark cage has been eliminated. The aft room will be used as a diving station with lockers, diving gear, hot showers, and an open hatch for access to the sea. The forward room will be used as an observation and storage compartment. It is fitted with large viewing ports, a refrigerator-freezer unit, and an emergency exit hatch.

The surface support ship for Sealab III is a modified LSMR (Medium Rocket Landing Ship). It will be held in a tight five-point moor so that there can be very little lateral movement, even in rough seas. For additional safety, the umbilical lines are rigged with counterweights so that the movement of the support vessel will have little effect on the habitat.

The divers will not swim down to the habitat. Instead they will be transported up and down in a kind of pressurized seagoing elevator called Personnel Transfer Capsule (PTC). Upon completion of the dive, the men will be brought up in the same way, and the PTC will mate with a Deck Decompression Chamber (DDC).

Once again human performance will be studied, specifically designed underwater tools and equipment will be tested, and further studies will be made of topography, geology, and marine life.

The men will be tethered during all work dives with air supplied from the habitat. They will, however, wear the new and improved Mark VIII semiclosed breathing apparatus, which will be used only in an emergency.

The water clarity and temperatures off San Clemente Island, California, the site of the Sealab III test, are considered better than those at La Jolla.

Besides the forty-five aquanauts, the Sealab III team includes two porpoises, two sea lions, and a harbor seal. They are trained for search and rescue of "lost" aquanauts and for the delivery of tools, emergency medical supplies, specimens for laboratory analysis, and other support items.

The harbor seal, who is a newcomer to the team, learns rapidly, can stay underwater at least twenty minutes without returning to the surface for air, and can make deep dives.

All the mammals can breathe the atmosphere of the habitat and return to sea-level atmosphere with no ill effects.

Aquanauts are very carefully selected according to a large number of criteria, psychological as well as physical. They are appraised by each other as well, since the final choice means they will be isolated and confined under difficult conditions as teammates for an extended period of time.

Their training is continuous over a period of at least six months, and it takes place in a number of different locations throughout the United States. It includes physical conditioning, a number of deep-water and shallow-water dives, simulated submersions, and long hours in the study of all the equipment they will be using. The ultimate goal is for the aquanaut to be completely free of breathing lines and able to find his way back to the habitat while performing chores away from it, and for the habitat to be self-supporting and free of any umbilicals to the surface.

This goal will probably be attained by the mid-1970's.

There is no question that we are slowly but certainly developing technologies and abilities to perform useful tasks on and under the sea. Almost simultaneously these developments are being utilized to exploit the economic potential of the sea's resources in such areas as agriculture, mining, and even recreation.

On February 15, 1969, two very different underwater habitats were lowered into very different seas, many miles apart, to begin a simultaneous experiment. One was marked for success, the other for tragedy.

On this date four civilian oceanographers began a sixty-day isolation in the new twin-capsule habitat called Tektite I. Designed and built by the General Electric Corporation, the 18-foot-high, 12-foot-diameter, 160-ton steel structure was lowered into 42 feet of water in Lameshur Bay just off St. John in the Virgin Islands. The test was to determine how well man can function after sixty days of isolation underwater. During this time the scientist-aquanauts were to conduct marine biological and geological surveys. Participating agencies in the project included the U.S. Navy, Department of the Interior, NASA, and the University of Pennsylvania.

Also on February 15, the Sealab III habitat, painted bright yellow and resembling a wheelless locomotive of the future, was plunged into the gray choppy waters off San Clemente Island, Southern California. It was done without incident. But two days later a man was dead!

Two slight leaks had developed in the habitat and experienced divers were checking. One of the men was Barry Louis Cannon, age 33, a civilian engineer from Panama City, Florida, and a veteran of Sealab II. Mr. Cannon for no apparent reason was stricken with a cardiac arrest. Warrant Officer Robert Barth swam the dying man back to the waiting PTC (Personnel Transfer Capsule), which immediately began its ascent to the support ship. Although heart massage was administered all the way, Barry Cannon was dead on arrival.

The Sealab III test was halted indefinitely. I am sorry for the death of a brave man, but I most strongly urge that the tests continue soon. Otherwise the life of Barry·Cannon will have been wasted.

6 · *Farming the Sea*

There are two sharply divided and contrasting opinions as to whether or not the sea can provide enough food to support the rising world population. There are a few who contend that the problem can be solved by increased agricultural productivity right here on our own terra firma. They point out that we have produced so much wheat that we store the surplus in bins. We dump whole crops of fruit rather then let the price go down.

All true. But try distributing this food throughout the under-privileged, overpopulated countries, and it will be absorbed without a trace and without alleviating the starvation problem.

During a formal address, former Secretary of Agriculture Orville Freeman admitted defeat in the United States's effort to close what he described as the "feedability gap" (ratio of food production to world population increase), with 150 million tons of American foods distributed to 100 countries since 1954 at a cost of 20 billion dollars. At present estimates, there should be enough food to sustain world population until 1980, but unless the "feedability gap" is closed, there is every chance that the world may run out of food by the year 2000.

There are knowledgeable people today who tell us reassuringly that in the next twenty years our food production can be doubled. That is very probably true, but if the world population continues to rise at present rates, the increase will be much more than double. Food increase will not be commensurate with population increase.

There are certain things which can and must be accomplished. Every foot of tillable, arable soil must become productive; we should develop extensive hothouse agriculture; we should process high-protein concentrates from every available source; and we must learn to exploit and increase the existing resources of the sea, namely plankton and fish.

The present-day world population demand for animal protein to sustain normal health is about 29 billion pounds per year. By the year 2000 it is possible that the world population may require 100 billion pounds. Utilizing current total land agricultural potential (including land unproductive due only to lack of proper machinery and knowledge), there is not the faintest hope that we can meet such a demand unless we utilize the resources of the food-rich sea.

Every country who has seas on her borders, or lakes and streams within her boundaries, should assume the responsibility of harvesting these areas to the utmost so that peoples of less fortunate countries can live useful, healthful lives. The resources and potential do exist to feed that "half the world population that goes to bed hungry tonight."

Some countries have adjacent waters but cannot afford the programs necessary to utilize them in the production of food and food concentrates. They must be helped in whatever way possible.

Most of all, the United States as a world power must set an example by fully utilizing the potential of the two greatest oceans in the world, which border her shores. By so doing not only can we fulfill our moral obligation to the underprivileged nations of the world, but we can increase the economic health of our own country.

Up to this point, exploitation of the food resources of the sea had little or no priority. There was more immediate profit to be made from the sea's mineral resources. Gradually this is changing, and more effective programs are being put into operation that will result in increased production of food from the sea.

In order to better visualize this food potential we will attempt at this time to present a comprehensive picture of world fisheries as they were some years ago and as they are today.

In 1965 the world fish catch contained nearly 20 billion pounds of protein. This figure is based on the fact that the net weight of fish contains 15 percent protein. This amount would have supplied approximately twelve grams (.420 ounces) of protein per day for one year to 2 billion people and alleviated chronic protein deficiency for that many individuals in the poor equatorial zones. This food, however, was not distributed so that protein-deficient people received their healthful share.

Although there are a number of imponderables, we can safely make the statement that we have not really begun to make full use of the sea's food potential. In order to strike a healthful balance of the population versus food, we must accomplish a more effective use of the sea and we must curb, through education, the sharp rise in the world population.

The total 1968 federal budget for development of fisheries and seafood technology was just under 50 million dollars. In 1966 that same budget was 38.7 million dollars. The increase shows awareness by the federal government, but the figures are still unrealistic compared to the size and urgency of the task at hand. If, for example, the budget for the development of the sea's resources were comparable to that for outer space development, we would do much to increase the economic potential of this country and the world.

It is estimated that in excess of 400 billion tons (and I believe that is a low figure) of organic material is produced annually in the sea. Man harvests but a tiny fraction of this.

In the last fifteen years the world fish catch has gone from 60.5 billion pounds to approximately 135 billion pounds. The United States places a bad fifth in the annual fish catch race, with Peru, Japan, China, and Russia well ahead. In 1954 we were second, with only Japan ahead of us.

America's annual fish catch is about 6 billion pounds, with the sport-fishing catch roughly 620 million pounds.

Peru today has an annual catch of about 28 billion pounds, with Russia at 11 billion and Japan at close to 16 billion pounds.

There is no legitimate excuse for a country with oceans on both borders, uncountable lakes, great wealth, technical knowledge, and almost limitless materiel, to be in fifth place in an area of great economic importance. Yet that is where we are, and the reason is clear neglect of a great economic potential that has always existed right at our doorstep.

The number of ships, tonnage, and equipment of our fishing fleet has lagged way behind that of the leaders. Our fleet has become too old and too small, with obsolete fish-finding equipment. We have not utilized fish resources right off our own coasts. For example: there is a standing crop of about 15 billion pounds of hake and anchovy in the California Current off the coasts of California, Washington, and Oregon. One reason for not harvesting this potential is because anchovy is important to the sport-fishing population. An agreement was finally worked out with the California Fish and Game Commission to allow an annual harvest of 150 million pounds. The actual potential of hake and anchovy is about 3 billion pounds, which could be made into protein concentrate.

Japan is the undisputed leader in the field of farming the restricted areas of the sea, such as bays and estuaries. The word that best describes this unique type of farming is "aquiculture."

The following describes the beneficial results of aquiculture methods.

Earlier in this book we described a possible method of creating fish and plankton "farms" by causing a mechanical upwelling of nutrients from the sea bottom. The Scots have fertilized some of their lochs, and the resulting fish population has been eighteen times as large as normal. The Japanese grow oysters from long ropes hanging from floating rafts. This method increased their yield from 600 pounds to 32,000 pounds per acre of seabed.

The United States Public Health Service has identified more than ten billion acres of seabed that are suitable for shellfish production. If aquiculture methods are used, this area (barring pollution) has a potential of 90 billion pounds a year.

Squid, shrimp, and crabs are cultured in Japanese laboratories until they reach a specified size; then they are allowed to grow under controlled conditions in natural waters.

Realizing this, the United States has encouraged the training of ecologists and has provided them with facilities and a program to increase the growth of shellfish. It is a beginning that needs to be watched and stimulated.

Just thirty years ago, the now-famous king crab found off the coast of Alaska was virtually unknown. In 1935 there were seven Japanese cannery ships, along with a fleet of small fishing boats, just north of the Alaska Peninsula in the outer fringe of Bristol Bay. It was thought that they were there to help themselves to the rich red salmon runs. A hue and cry went up. A stringent protest was made by Secretary of State Cordell Hull, and the Japanese fishing fleet obligingly withdrew. In the course of the investigation it was discovered that the sea-wise Japanese were not after the salmon at all. They were after the huge, succulent king crab.

In a very few years king-crab fishing became Alaska's largest industry, but it had yet to meet a new threat.

In 1959 both the Russians and the Japanese had ship canneries and fishing fleets off Alaskan waters. In four years the foreign catch of king crabs skyrocketed from 2 to 9 million pounds. The United States was in imminent danger of being fished out of its own grounds.

Fortunately, in 1958 the International Conference on the Law of the Sea in Geneva "assigned living organisms in constant contact with the seabed to the adjacent coastal nation." Negotiations were made on the basis of this allocation. Catch quotas were greatly reduced, foreign vessels were limited to certain fishing areas, and mutual conservation laws were enacted.

We settled the Alaskan crab controversy because of the fortuitous existence of a law. The fact remains that we should have known the king crabs were there, and we should have known how to fish for them.

The American fishing industry operates from its own coastal fishing grounds, which are increasingly being invaded by foreign fishing fleets from Russia, Japan, and Europe. Russian trawlers ply American coastal waters from Cape Cod to Florida. The Russians have larger vessels, better catching equipment, and more flexibility of operation, and they fish as completely self-contained units. The foreign fleets are more efficient because they process their catches *on board*, delivering the end products to their own shores ready for the markets.

The compilation of recent statistics points out that we have a mammoth fisheries potential which could result in a sustained yield of 22 billion pounds, instead of the present annual catch of slightly less than 5 billion pounds.

It comes as a severe shock, then, to realize that the United States is the largest fish-importing nation in the world. With fish catches declining, we are importing 50 percent of our fish consumption. It is most assuredly more costly to import fish than to catch it off our shores. Increasing our annual catch would do much toward strengthening the economic health of our economy. As a matter of fact, we have enough potential on our own home grounds to be in a position to export large quantities of fish and byproducts to other parts of the world, resulting in increased revenue for this country.

The federal government is stepping in with subsidies for more and larger fishing vessels. We will also need increased tariffs on foreign fish products, faster and more economical fish processing and handling methods, better methods for locating and fishing more prolific fishing grounds.

In this respect private industry has made recent valuable contributions. EG&G International, for example, has designed and built the superior quartz iodide incandescent light and photo system; Bendix has built acoustic telemetry systems; Sperry Rand has developed a deep submergence viewing system; Interstate Electronics Corporation, Ocean Systems, Westinghouse, IBM, Ogden Technology Laboratories, Corning Glass, and General Dynamics—all are making invaluable contributions toward the understanding and knowledge of the ocean bottom. It is only with better knowledge of the ocean environment that we can develop our sea's resources.

Therma-climes, bottom topography, current, waves, salinity, temperature changes, and plankton—all contribute to the migration and concentration of fish, one of the most nutritionally perfect foods known to man. It is these ingredients that are being studied, and about which data are being compiled, through the use of more and more sophisticated equipment invented and developed by private industry.

Ironically, even among the individuals most involved with the man-in-the-sea program, the potential of fish protein concentrates is a bone of contention. In a way this is explainable. Private industry must always present its stockholders with a profit and loss statement at the end of the fiscal year. Consequently, its underseas programs must bring quick profits. The moral obligation of feeding the world population is not the primary concern. Thus, instead of getting involved in such things as the development of fish protein concentrates, industry tends toward other, more profitable areas, such as the extraction of petroleum from offshore deposits.

Nevertheless, applied research by the Bureau of Fisheries has come up with a successful, relatively simple process for extracting low-cost animal protein from the lesser-used fish of the sea, such as hake and similar species. This protein can be added to cereal products at a five to ten percent level, without adding a fishy flavor. Ten grams can provide adequate animal protein for one child at an estimated daily cost of less than a penny.

Congress has authorized the construction of a pilot plant, and the Food and Drug Administration has approved PFC (fish protein concentrate). It is proposed next to survey the local fishing potential and marketing ability of three selected, less developed countries. The initial goal would be for these countries to provide, by 1971, ten grams of animal protein daily to as many individuals as the experiment would allow.

There is little doubt that specific species of fish have dwindled or even virtually disappeared from certain areas. Up to the present time the mass exodus from their familiar geographic locations has been a mystery. The veteran fisherman merely says (and believes fervently) that ye olde fishing grounds have been "fished out." As a diver of some eighteen years' experience, I very seriously doubt this.

As a matter of fact, over the years I have been hired a number of times to go down and check areas that have been allegedly fished out. I never, on these occasions, found a fishless area. What I did find was a smaller number of more wily fish, hiding in every conceivable place and too cunning to be caught. In almost all cases these survivors of intensive fishing operations were fairly large specimens, both male and female. I advised a system of crop rotation that would allow the area to lie fallow for a short time. The fish population quickly built itself up to normal levels. In other cases, I found that the fish had migrated to other areas, with propagation taking place in the midst of flight.

There have, however, been great reductions in some fish populations. The menhaden, for example, have dropped from a 1961 catch of over 2 billion pounds to a current catch of 1 billion.

Now the menhaden is a prolific fish that can drop 700,000 eggs with one wiggle and two shrugs of its tail. It is not sought after as a delectable morsel for the table, but it is coveted by commercial fishermen for its rich oil content, as well as its meat and bones, which can be ground into a high-protein animal and poultry feed. The oil can be used in all kinds of paint from cosmetic (lipstick) to house paint.

There are a number of possible reasons for the drastic reduction in menhaden catches in certain circumscribed areas. The fishing boats operate with purse nets that are 1,200 feet long and 60 feet deep, which, quite naturally, bring in tremendous catches. Worse than that, boats have been operating in the spawning fields of Chesapeake Bay. This is a good way to obliterate a species from the face of the earth.

I suggest that the fishing fleets seek other pastures for a year or two and give the menhaden a chance to regenerate themselves.

Another method of helping sustain and even increase the fish population is through the use of artificial reefs. This idea was launched successfully in Florida in 1960, by the Englewood Fishing Club, which sank off its shores sixty wrecked car bodies, twelve stoves, forty refrigerators, and a potpourri of other junk. The next year the members were hauling in cobia, grouper, and jewfish. A reef of large, hollow concrete blocks was constructed at Sand Dollar Ridge, two and a half miles off Cocoa Beach, where the fishermen are now bringing up red snapper, bluefish, sea bass, whiting, kingfish, grouper, and tripletail.

So save your old ice boxes, refrigerators, tires and stoves, and help build artificial reefs. As a diver I have seen a great deal of junk under the sea. The minute organisms of the water have a way of changing its shape into something quite acceptable, and I can guarantee that it will have immediate occupancy.

While still in the mood for the oddity, I would like to note here that the lowly barnacle is being studied for possible application in dental work. The barnacle secretes a very strong natural cement, which might be used one day to retain fillings in cavities. You may never again have to worry about eating corn on the cob.

The Alcoa Company has come up with a lights-and-camera setup for fishing nets. The results will be monitored in the vessel above, allowing the captain to study the escape routes of the fish.

In pursuit of a more pragmatic study of fish population recently I journeyed to Yarmouth, Nova Scotia, and hired myself out as a crew member on one of the lobster-fishing boats. It was a bitter, raw, gray November day as we left the pier at 8:15 in the morning, the first day of the lobster season. Every available foot of deck space was piled high with lobster traps. I had to sit on the gunwales and let my feet dangle over the side. A sadistic wind whipped through my clothes and chilled my bones. As I tried to huddle within myself, I wondered sadly what strange impulse had led me to this cold and miserable perch. Of course I knew what it was. I wanted to know exactly what it felt like to be a lobster fisherman. My musings were cut short by an order from our captain. It was a terse alert to get ready with the traps.

When we were over the lobster grounds we heaved the traps and weights over the side. We had an allotted time to get as many traps in the water as possible.

In just one hour of backbreaking work we had every trap in the water, and I sat down on the afterdeck, leaning my back against the stern. I was sitting in all the fluid and oil from the fish bait that goes into the traps. It stank—in fact, it was the worst thing I had ever smelled—but I didn't care. It was a slice of heaven just to sit there.

The respite was short-lived, because our good Captain Bligh had banked his boat in a turn and headed back to the dock. The work had just begun. There were more traps waiting for us on the docks. All the traps were color-keyed so the crew would know which belonged to the various boats.

We were loaded and back out on the choppy waters within

fifteen minutes, heaving the heavy traps overboard as soon as we hit the lobster grounds.

At about 1:00 in the afternoon a tragedy struck. One of the youngest members of the crew had allowed his foot to become entangled in the lines and had gone overboard with the trap and weight. Without thinking I jumped in, reached out my arms, and groped for him. I felt nothing except the crushing cold of the water. I couldn't breathe. I could just barely move. Hands grabbed me from behind and hauled me back into the boat. The captain quickly maneuvered back to where the other man had gone in. The other men rapidly hauled the trap and lines aboard. The young man was still entangled, but he was dead.

Accidents such as this are not numerous, but serious accidents do happen nearly every season. They are part of the fisherman's way of life.

The lobstermen are not allowed to fish their traps on the same day they place them. They must wait until after midnight. At the time I was on a crew strong winds and heavy seas kept the boats in the area shoreside until daylight of the following day. I had had enough as a crewman, but I stayed around for two more days to collect data on the various catches.

Each boat was bringing in between 700 and 1,000 pounds per catch. With the price set at 75 cents a pound, some of the captains were making up to $750 a day. There are two six-week lobster seasons, so with luck a hard-working crew might split between $60,000 and $80,000 a year. An average boat crew is about four men.

In Nova Scotia the price of the lowly scallop has gone up to the same price as lobster—75 cents a pound. I checked on one scallop dragger at the Highliner fish plant who had off-loaded a catch of 24,000 pounds. That is $17,000 for one catch. Scallops are a risky business, however, since just one year before the price per pound had been 34 cents.

Another rich harvest Nova Scotia can claim is pollock. One vessel, the *Lady Anna*, brought in a record 207,000 pounds. The *Lady Anna* is a converted 100-foot midwater trawler, which had formerly been a scallop dragger.

Another region that is making the most of her natural marine resources is the Republic of South Africa. She has become, in very recent years, the largest producer of fish and fish products in the Southern Hemisphere. Together with South-West Africa, this still somewhat undeveloped area now ranks sixth among the world's fish producers. South Africa is, by herself, the world's largest producer of fish meal.

Her total annual fish catch increased from 190,000 tons in 1948 to 1,450,000 tons in 1968. Approximately 90 percent of this catch is exported to the United States, France, Holland, the United Kingdom, and Australia.

So the potential is there. The sea is food-rich. To utilize the potential fully, our federal government must prepare more realistic budgets. We need federal agencies that are structured and empowered to activate expeditions and forceful programs. We have a committee called The National Council on Marine Resources and Engineering Development. This council made a long, exhaustive study and discovered that we, as a world power, were somewhat behind in several of the marine sciences.

All well and good. We have had the study. Now we need a marine resources committee that can lobby for larger and more realistic budgets. The membership should be made up of men and women with a solid understanding of the marine sciences.

Back to the question. Can we feed that half of the world population that goes to bed hungry tonight?

Yes, it is possible, but ultimate solution is still a dishearteningly long way off. The world program must be implemented by education in birth control and the utilization of all agricultural areas, especially that of the sea.

Man has managed to plunder the treasures of the sea for other benefits with a much larger degree of success. Specifically, he has done well in mining the seas. Most of the precious metals have succeeded in evading his efforts, but he has managed to extract petroleum in great quantities from offshore wells.

7 · Mining the World's Waters

The continental shelves of the world are actually submerged extensions of our known geographic land masses. They constitute an area of about 10 million square miles, which is an additional 20 percent of the above-water portions. Since the shelves are merely flooded portions of the continents, it follows that their mineral properties should be similar to those of the land. This is proving to be the case in a number of areas.

Important offshore oil deposits have been discovered and are already successfully productive in a number of places, especially in the Gulf of Mexico and off the coast of California.

Placer deposits of gold and platinum have been discovered in the submerged beaches off Nome, Alaska.

Tin is mined from river placers in Malaysia, Indonesia, and Thailand.

South Africa has always been rich in diamonds. Now these precious, sought-after gems are being found in such places as the Orange River in southwest Africa and certain locations in the Amazon River.

Phosphorite deposits have been found in substantial measures off the coasts of Chile, Mexico, Peru, Argentina, South Africa, Japan, certain islands of the Indian Ocean, and both coasts of the United States. Some of these deposits are estimated to be 96 miles long and at least (according to seismic probe findings) 65 feet deep.

Coal and iron are being mined from seabed rocks in England, Newfoundland, Japan, and Finland.

One of the less exotic mineral commodities presently being mined from the shelves is sand and gravel. It is of great economic value, nevertheless, since the quantities are copious and it can be used in a number of ways, such as construction, land reclamation, and the rebuilding of eroded beaches.

Getting into the deep waters, we find that manganese nodules are the most interesting of the ocean floor sediments. These valuable objects are very small (averaging 2/5 of an inch in diameter) and friable (crumbling easily). These nodules are 36 to 50 percent manganese and contain smaller deposits of copper, nickel, and cobalt. They are brown to black in coloration and lie in vast, thick deposits throughout the deep oceans.

Estimates by such knowledgeable people as John L. Mero of Ocean Resources, Inc., indicate that the Pacific Ocean floor alone holds some 1.5 trillion tons of manganese nodules and that they are forming at the annual rate of 10 million tons. In some areas there may very well be concentrations of 100,000 tons per square mile.

One serious problem exists, unfortunately. Most of the dense concentrations lie in extremely deep water. During the International Geophysical Year program, Russian deep-water cameras photographed nodule beds at 18,000 feet. Not all concentrations are that deep, but even in slightly shallower depths of the abyssal plains we are totally inept. We do not have the equipment to successfully mine these areas. This does not by any means suggest that we should allow this area to remain latent. Federal budgets should be allocated and technology should be developed which will one day facilitate deep-ocean mining. One of the first areas for exploration and extraction of manganese nodules should be the Blake Plateau off the Florida coast.

The same holds true of other precious and coveted metals of the sea, such as gold, radium, and silver. We are several decades away from being able to extract these metals from seawater in any practical manner.

We should consequently concentrate the main part of our efforts in the more accessible waters of the continental shelf.

Under the International Convention on the Continental Shelf, the United States and other sea-bordered countries obtained sovereign authority over their adjacent continental shelves and slopes, as well as the resources they contained, to the greatest depth at which they may be effectively exploited. This gives us an immense new territory of at least 850,000 square miles. Deposits of petroleum, natural gas, and sulfur have already begun to be extensively exploited, yet only a small percentage of the possible areas are in active production.

To date the petroleum industry has drilled approximately 9,000 offshore wells in the coastal waters of the United States, mostly in the Gulf of Mexico and off California. Close to 7 billion dollars have been invested in the installations, and 4 billion dollars have been paid to federal and state treasuries. The latter is for federal and state offshore leases that allow companies to explore and exploit the deposits.

More than 2 billion barrels of petroleum liquids and 6 trillion cubic feet of natural gas have already been taken from the continental shelf. An additional 3 billion barrels of petroleum liquid and 27 trillion cubic feet of gas await drilling in known reservoirs, and production will be well underway in these areas by the end of 1969. This is only a beginning. Although drilling is always a risky proposition as far as investment is concerned, there is little doubt that this industry will not only sustain itself but go on to greater and greater efforts.

Private industry is more than willing to invest large sums of money in this area of oceanography, for the obvious reason that the monetary return for investment is more immediate and of greater proportions than it is for other projects. This may not be as altruistic as an applied effort to feed world hunger, but it is sound economy. Industry is in business to make a profit, not to collect plenary indulgences.

The continental shelf is generally considered to extend to a depth of 100 fathoms, or 600 feet. The end of this shelf is usually marked by a sharp increase in the downward slope. This is called the "shelf edge." The incline slants steeply down to about 1,000 fathoms and is known as the "continental slope." The ocean bottom is referred to as the "abyssal plain," except for the deepwater trenches, or canyons. The average depth of the ocean is something over 12,000 feet, and the deepest trench ever explored by man is the Marianas Trench in the South Pacific, which is 37,800 feet deep.

We have said that the continental shelf gives the United States an additional 850,000 square miles of submerged land and its potential resources. Recent and continuing developments in technology have placed us at the threshold of another acquisition of flooded land. The continental slope gives us an area of approximately half a million square miles of petroleum and mineral potential, although at this date we have not explored the area enough to estimate this potential accurately.

The estimates of potential petroleum that may be extracted from the combined areas of both the continental shelf and slope vary drastically. We will learn much more in the next few years, providing, of course, that the federal government channels more funds into oceanography. Nevertheless, I would like to make an estimate of the petroleum potential of these combined areas that I do not consider to be overly optimistic.

The *world's* offshore petroleum potential should be at least 1,500 billion barrels of petroleum liquids and 2,000 trillion cubic feet of natural gas.

In the offshore waters of the United States, the areas of greatest possibility for productive wells are off California, Alabama, Florida, Mississippi, and possibly Washington and Oregon, and in the Gulf of Mexico, the Southern Bering Sea, the Gulf of Alaska, and Cook Inlet.

There are two types of platforms for offshore drilling—the fixed platform, which has not been able to withstand the massive power generated in sea storms, and the floating platform, which is more mobile and has been able to ride out heavy seas created by storms. Inherent to the floating platform was the problem of holding a stable drilling position, but that has improved with the development of dynamic anchoring and positioning. The next and perhaps ultimate step in offshore oil drilling platforms will be the completely submerged platform. This will eliminate any problems created by violent storms. The platforms operating well under the surface will feel none of the turbulence.

The birth of marine drilling came in the early 1930's in about six feet of water, in the bayous of Louisiana and Texas. The industry's engineering technique developed rapidly, and by 1965 the Shell Oil Company was drilling off the coast of California in 574 feet of water, using a floating semisubmersible rig.

In March 1962 a federal oil-lease sale was held off Louisiana and Texas. Close to two million acres were leased to thirty different oil companies. Of this total 820,000 acres were in water up to 200 feet, 213,800 acres were in water up to 300 feet, and 37,000 acres were in water up to 600 feet. All the leases had a primary time limit of five years in which to establish actual production. At that date there were only twelve floating drilling rigs existing in the world.

Private industry came to the rescue, and today there are more than fifty floating rigs, plus other sophisticated supplementary equipment. Thus the leases will achieve full potential yield in the five-year time limit, and the project will become economically profitable.

EG&G International, for example, has developed a seismic profiling system which draws electronic graphs or "pictures" of the ocean bottom with remarkable detail. This same system can be utilized with equal success for marine geology, archaeology, submarine cable and pipeline surveys, and harbor development.

Ocean Systems, Inc., developed three pieces of deep diving equipment which are proving to be of invaluable help to investors in offshore wells.

One development is known as the ADS IV (Advanced Diving System IV). It consists of three units—a submersible decompression chamber and two deck decompression chambers. The submersible chamber is a sphere with a number of viewing ports and is large enough to house two divers. It is used merely for the purpose of transporting the divers to and from underwater work sites. Immediately upon being hoisted topside the pressurized chamber is mated and sealed to an entry lock common to two decompression chambers. In the decompression chamber the divers can rest at saturation pressure (the pressure at which they worked underwater) and wait to be returned to the job site while another team takes over, or if the job is completed they merely serve their regular decompression time. This ADS IV is designed for operating at depths up to 1,000 feet.

In August 1967 two divers, Arthur Pachette and Glen Taylor, performed maintenance work on an oil well head at a depth of 636 feet, at a project site forty miles off Grand Isle, Louisiana. In this historic and record-breaking dive, the two men returned to the bottom of the Gulf of Mexico twice in order to complete their work. They were then returned to the surface and entered the deck chamber for the interminable, monotonous, very gradual decompression time of six days.

A second revelation in the realm of *in situ* (in place) underwater repair is a new underwater welding technique developed by Union Carbide's Linde Division and Ocean Systems, Inc.

In a recent demonstration a Tig (tungsten–inert gas) weld was accomplished for the Humble Pipe Line Company: a six-inch connection was tied to a ten-inch pipe line at the bottom of the Gulf of Mexico in more than a hundred feet of water. Until this time a Tig weld, which is the strongest kind of joining, had only been accomplished topside.

To do the job, a small work chamber was lowered and secured in place, directly over the pipe. The chamber was made the same pressure as the ambient water; it provided a *dry* atmosphere in which the welder could work. The ADS IV pressure chamber was lowered to a spot just a few feet away from the work chamber. Two men with scuba-diving equipment slipped out of the ADS chamber, swam the few intervening feet, ducked under a hatch, and found themselves in a small but dry work shed under a hundred feet of water. One of the men, a Tig welding expert, was not a professional diver, but he had received a week's instruction. An experienced diver accompanied him and assisted in what has become the forerunner of *in situ* platform underwater repair. Until this time, it had been impossible to perform an underwater weld comparable to the superior kind of welding that can be done on the surface.

In March 1968 the small submarine *Deep Diver*, which is the unique lookout vehicle described in the chapter on modern submersibles, settled on the floor of the seabed somewhere in the Bahamas in 700 feet of water. Two divers, Roger Cook and Danny Breeze, exited the chamber and for fifteen minutes collected various specimens of marine flora and fauna. Since it was quite dark, the exterior lighting system of the *Deep Diver* was utilized. The divers noticed an unexpected current, which was theorized to be an eddy of the Gulf Stream. The current was fast enough for the pilot, George Bezaak, to find it necessary to reposition the submersible during the dive.

After the divers completed their mission they reentered the chamber and began a scheduled twenty-five-hour decompression period. The divers reported that they felt no different during this deep dive than they had at shallower dives of 200 and 300 feet.

This proved to be another seven-league stride forward toward providing *in situ* platforms for important underwater missions.

Although petroleum products appear at present to be the most profitable of the underseas resources being developed, others are increasingly evident.

An important source of fresh water was discovered in 1966 by the Lamont Geographical Laboratory while drilling into Florida's Blake Plateau. (This operation, incidentally, was part of a government-supported program.)

The conversion of salt water into fresh water by desalination, although not undergoing any startling breakthrough, has been improved by a series of processes, such as multistage flash distillation, which are gradually lowering the cost of saltwater conversion.

The British very recently discovered a method of extracting uranium from seawater.

Although it has been assumed for many, many years that the concentration of trace elements is uniform throughout the world's sea waters, it has been discovered through improved methods of water analysis that this is not necessarily the case. Certain water areas surrounded by hot deserts contain concentrations of iron, copper, and zinc up to 40,000 times that of average seawater.

Large concentrations of phosphates have been found in shallow water, from 100 to 200 feet in depth, which will make profitable mining possible within the next few years.

Alongside this pleasant and optimistic prospect hovers the long, rather ominous shadow of various legal questions. Who owns the mineral and food wealth of the continental shelves of the world? What are the geographic limits of the shelves? As underseas exploitation extends into deeper waters, who shall possess the rights to the territorial seas? What geographic delineation shall be made of these waters? Will the seeking of equitable answers bring the various countries involved even closer to the threat of war than the economic and political ideologies of today? Or will the situation, by the grace of a beneficent God, bring us closer together?

I am happy to report that judging from the first negotiations, we are striving toward a fair accord. The discussions, although expressing widely divergent opinions, were accomplished in an open-minded and friendly atmosphere.

This does not mean that the situation couldn't end in one of the most frenetic land grabs (considering the continental shelf and slope as land) of contemporary times. This is something we must avoid at all costs, and that can only be done by complete resolution of all international legal questions.

During the 1967 Session of the General Assembly of the United Nations an item arose concerning the resources of the seabed that was known as the "Maltese Resolution." It was so named because its introduction came from Dr. Prado, the Maltese representative to the United Nations. The intent of the resolution was to do for the seabed what has already been done relative to the intents and purposes in the exploration and utilization of outer space.

Dr. Prado proposed that the seabed, or inner space, should never be used for the narrow purposes of national interest, whether military, economic, or territorial. Every effort should be made to ensure its use for the benefit of mankind as a whole, for undeveloped nations as well as world powers. Finally, it should, as far as possible, be demilitarized.

Unfortunately, the Maltese Resolution aroused very little comment or interest from the local or international press, but it did create one of the liveliest discussions the United Nations has witnessed for some time. The delegates all seemed to realize that grave problems could and would arise unless effective action were taken.

I would like to describe some of the problems that exist right now regarding this issue.

In 1958 the International Conference on the Law of the Sea reached an agreement concerning the continental shelves and their territorial jurisdictions. This agreement stipulated that a continental shelf extended to a depth of 200 meters (660 feet) or "beyond that limit, to where the depth of the superadjacent waters admits of the exploitation of the natural resources of the said areas." The latter part weakened the law to a point where it is almost valueless, because it leaves itself open to the individual interpretations of various nations. With the recent development of underseas technology, the present "ability" to exploit the natural seabed resources virtually wipes out geographic limitations. The "boundaries" could stretch across the oceans from continent to continent. And what of islands? They, or the nations who own them, could claim a substantial portion of the ocean waters.

Stringent international laws regarding pollution of the seas must also be enacted immediately. Not only oil refuse but pollution by radioactive wastes, which can contaminate the entire environment under the sea, must be controlled.

It is my firm belief also that inshore nations and states—not only the coastal areas—should share in the wealth of the seas.

During the 1958 maritime convention twenty-one nations, including the United States, favored the three-mile limit to the territorial sea; seventeen nations favored a six-mile limit; the Soviet Union and ten others wanted twelve miles, and five nations wanted 200 miles.

The Second Conference on the Law of the Sea convened in Geneva in March 1960 and once again tried to fix the limitation. The United States and Canada proposed a six-mile limit, while the Soviet Union and seventeen other nations stuck to their proposal of a twelve-mile limit. Once again the situation was unresolved, and the territorial limits are still legally regarded to be at three miles. There is no question, however, that this limitation will be extended.

Finally, the subject of demilitarization must become one of swift action. We are rapidly approaching a time when a nation will be able to establish missile bases underwater. We already have submarines that can fire a missile while submerged; the vessel itself can stay submerged for months at a time. The oceans of the world could become a battleground. We can and must prevent this, by international agreement.

In July 1968 I journeyed to London for meetings with members of Parliament and British private industry. The atmosphere there was more serious and farsighted than that of other countries to which I traveled. In my search for answers I visited England, Germany, France, Italy, and other European countries.

The most realistic thinking was summed up by Patrick Armstrong, Honorable Secretary of the Parliament for World Government of the Sea's Resources. He pointed out one fact with succinct directness: Unless we organize a world governing body to regulate and supervise all activity and legal aspects related to the sea's resources, there will be a land grab like none other in the history of man. Worldwide anarchy could easily result.

There is more than enough wealth in the waters of the world for everyone—for all their comforts and needs.

A world organization *must* be established so that the ocean's resources will be used for the benefit of all mankind!

Before I left London I offered to represent America on this vital committee—Parliament for World Government of the Sea's Resources.

8 · A General History of Man Under the Sea

There are still a few people today who labor under the delusion that skin diving and scuba diving are new, offbeat sports kept alive entirely by the gung-ho, leather-jacketed, rock-and-roll crowd. Nothing could be farther from the truth. The rapidly growing diving fraternity contains a cross section of all occupations in its membership.

Let us first go into a quick history of this "new" sport. Naked divers penetrated the Mediterranean Sea untold ages ago. Mother-of-pearl, which cannot be gathered in any quantity without diving for the shells, has been found in carved ornaments from excavations that date back to about 3200 B.C.

Aristotle, the first scientific naturalist, wrote so accurately about fish that one can believe he was a diver.

Xerxes employed combat divers, but where he acquired trained divers we do not know.

It is not even necessary to examine these scraps of information. The evidence of ancient diving is obvious in the marine products used by the ancient Greeks, which came from creatures that were attached to the bottom of the sea and could be harvested only by divers. Imperial purple dye, Caesar's favorite color, came from a shellfish. Sponges were widely used by the Greeks. Roman soldiers soaked them with water and carried them as canteens. Divers working in choppy waters held oil-soaked sponges in their mouths. They would bite the sponges and send up oil to lie on the surface, cutting down the bothersome beams of sunlight. Roman mothers soaked sponges with honey to give to their crying offspring.

The much sought-after Golden Fleece may actually have been the long, silky byssus of the large pinna clam. Shells of the giant tridacna clam were used as holy waterfonts in medieval cathedrals.

In the year 1000, when the Vikings were at the height of their fame, there was a famous Danish pirate named Oddo. This man was such a marvelously skilled seaman that he developed a reputation of almost being in league with the devil. The king of Sweden chose one of his lieutenants, Eric the Eloquent, for the dangerous honor of fighting Oddo. Now Eric was a brave man, but being also a smart one, he knew the danger of battling someone reputed to be in league with the devil, and so he set up a plan against sorcery. During the night he sent several of his hardy, deep-chested divers below the surface to drill holes in all of Oddo's vessels. The following morning, when Oddo's ships began to list and his crews were busy bailing out the water, he attacked. The Danes, distracted by the need to save their ships, were unable to withstand the assault.

Stories of diving appear in the chronicles of the Middle Ages, especially in those of naval warfare. Jehan de Bueil, a French admiral, was well up on the technique of diving, as witness his instructions for the mustering of "certain men well instructed in the art of diving and able to hold their breaths under water for a long time, so that during battle they may swim unseen towards the enemy with strong drills, to bore holes in the enemy hulls, thereby causing the water to enter as through a colander and causing the ship to sink."

Even sabotage commandos were known in the Middle Ages. For example, in the year 1203 King Philip Augustus of France was besieging the island of Andelys. A fabulous man named Gauberte, a trained diver, swimming underwater, carried combustible materials in small earthenware pots to the palisade, which formed the access to the island and which also housed the main defenses. Unseen, he ignited the material and burned the fortress. In 1372 Franco-Spanish forces, under the command of Charles V and Henri de Transtamare of Castille, surprised a number of English ships, loaded with gold, off La Rochelle. Towards the end of the battle the Spanish came up with a little surprise, which was quite rare for those days. They launched a number of innocent-looking small boats filled with wood that had been impregnated with inflammable liquids. These little floating bombs were manned by skilled divers trained in their use, and when the time came they were set alight and directed against the ships of the duke of Pembroke. This strategem took the battle-experienced English so much by surprise that thirteen of their large vessels were lost.

At about the same time a poet named Raimon Jordan also wrote of the same man. Towards the end of the year 1210 Gervais de Tilbury refers to a certain Nicholas, who is said to have asked various fishermen for oil so that he could more readily descend into the sea. One day, at the order of his king, he plunged into the famous whirlpool at Scylla, and on his safe return he described his reactions to the king and court. Numerous other writers have referred to the exploits of Nicholas, "the fish."

There is even a specific allusion to this man in Don Quixote: "Quite apart from the fact that the knight errant was vested with all the theological and cardinal virtues, and ignoring minor details, too, I must record he could swim every bit as well as could Nicholas the Fish."

At the end of the twelfth century an Englishman, Walter Mapes, described a certain Nicholas Pesce, "the diver." He was a man who was able, because of a life spent almost constantly in the water, to penetrate the secrets of the sea so completely that he could even predict storms. As an object of great interest, he was taken to the court of King William of Sicily, where, separated from that which had become his element, he gradually pined away and died.

The seventeenth century was still talking of this famous Nicholas. Two scholarly Jesuits, Father Fournier and Father Kircher, both refer to his exploits: "Nicholas, the Fish, swam so well against the current, even during the worst storms, and was able to remain under water such a long time, that some people, not realizing what long practice can do, did not believe such a thing was possible without the aid of magic."

The Crusades led to the swift development of the technique of sea warfare in the Mediterranean. Having conquered Cyprus, Richard Coeur-de-Lion left the island to join King Philip Augustus of France. On the way he encountered a large vessel flying the French king's flag. Richard immediately sent two galleys toward it with an invitation to the captain to visit him on board his own vessel. In reality the ship was full of Saracens, fifteen hundred of them, on their way to raise the siege of Acre. The deception was soon revealed, and Richard gave the order for the galleys to ram the enemy. According to Matthieu Paris, divers also took part in this operation, swimming underwater to the Saracen ship and boring holes in its hull in several places.

There is another account of the story by August Jal, specialist in the history of naval warfare. Jal believes that what the divers actually did was to jam the rudder of the Saracen ship, rendering it an easy prey to the Christian galleys. At any rate, the Crusaders scored another victory, due to the activity of divers.

The diving methods and equipment employed in the Middle

Ages were exactly the same as those used in antiquity: the free dive, the diving tube, and the diving bell. In the twelfth century the monk Roger Bacon wrote: "Devices can be made by means of which men could walk on the bed of the ocean without harm to their bodies. Alexander (the Great) used such a device to discover the secrets of the sea." But very real courage was still necessary for a man to dive down into the sea, which in popular belief, was still inhabited by many and terrible monsters. "The dogs of the sea," Olav Magnus informs us, "lie in wait hidden near ships at anchor in order to seize any sailors who may fall in. Divers protect themselves against these animals by carrying pointed stakes with them, so that they can kill them when they attack. Terrible battles take place underwater between divers and these beasts, which seek particularly to devour the genital organs, the feet, and other parts of the body which they regard as specially appetizing."

Another example of the use of diving in medieval warfare occurred in 1421, when the soldiers of Aragon were besieging the Genoese port of Bonifacio. They had sealed off the entrance to the port with thick piles and a heavy chain. Unfortunately for them, reinforcements arrived from the sea for the besieged Genoese. But what finally caused their defeat was the enterprise of a hardy Genoese diver who swam underwater to the ships of Aragon, armed with a cutting tool, and proceeded to cut all the anchor cables so that the ships drifted off at the mercy of the winds and tides.

In all the warm seas, people have been diving for thousands of years. Frederic Dumas found a seventeenth-century French print of red-coral divers wearing goggles. The goggles may have been made from polished transparent tortoise shell.

Wall paintings in Arnhemland, in aboriginal northern Australia, show Stone-Age men swimming with spears and strings of fish. Leonardo da Vinci sketched a number of different diving rigs in the fifteenth century. An eighteenth-century French engraving shows a free salvage diver wearing air bottles on his back. Man was not able to really begin exploration of the underwater world, however, until the nineteenth century, when technical advances were made that increased both the period of time he could stay beneath the waves and his mobility while he was there.

Applications for patents not so different from Da Vinci's early concepts began to appear in increasing numbers at the Patent Office in Washington, D.C. At the same time similar applications were made in other countries, such as Italy, France, England, and Germany.

Sam Davison, Jr., of the Dacor Corporation, has spent much time and money in becoming a kind of diving patent historian. Thanks to him, this book gives credit to the man who patented the *first* practical, independent diving lung. It was a perfect patent in all respects, calling for a compressed air cylinder and using a demand type of regulator. There is no doubt that this is the true father of our modern scuba (self-contained underwater breathing apparatus). The inventor to whom we refer is a Parisian named Benoist Rouquayrol. The date of his patent was November 6, 1866—many years before Jacques Cousteau developed the aqualung.

A few years before, in 1839, the world's first practical helmeted diving suit had been invented by August Siebe, who was the founder of one of the most famous of all diving equipment firms, Siebe, Gorman & Company, Ltd. One of the latest variations of the helmeted suit is the self-contained injector dress, invented by Sir Robert Henry Davis. Sir Robert was also an author, and the writers of today, myself included, owe him a debt of gratitude for his accurate and well-researched chronicles.

The first independent diving lung was fabricated by Henry Fleuss of Siebe, Gorman in 1879—thirteen years after Rouquayrol's patent. Fleuss was a wiry, lively, ex-merchant marine officer. He built an oxygen rebreathing device, whose carbon dioxide absorbent was a solution of caustic potash soaked in rope yarn.

Fleuss tested his device at Wooten Creek, Isle of Wight, in 1879. He was accompanied by several rowboats filled with well-wishers and deeply concerned friends. His diving gear consisted of a kind of flexible bag strapped to his back and connected to a copper tank. Tubes ran from this to a stiff rubber mask, which was fitted to his face. There were lead and iron weights attached to a belt around his waist, and metal chains were wrapped around his ankles. It looked as though an esoteric order of religious cultists had gathered at Wooten Creek to sacrifice one of their members to the Wooten water gods.

Henry Fleuss labored upright under his heavy diving gear and prepared to vacate the rowboat. One of his friends held him back, concerned about the excessive weight. Wisely he suggested tethering the eager diver to the boat.

Without waiting for an answer, the man tied a long, sturdy line to Fleuss's belt.

Henry Fleuss stood on a seat and dropped feet-first into the water. He plummeted down, scattering frightened fish in all directions. Landing heavily, but still on his feet, Fleuss looked around. Through a kind of green translucence he could see quite well. Fascinated, he forgot time and strolled about the river bottom, enraptured with each new sight.

In the boat above, his friends waited apprehensively for reassuring tugs from below before they reluctantly paid out more line. Checking their watches, they noted that Fleuss had been down somewhat longer than in the tests in the pool at the Old Polytechnic in Regent Street, London. Still, there was no sign of trouble, and so they bided their time.

It was a good thing that they couldn't see below, because Henry Fleuss was about to get into trouble. At first he was content to revel in the marvelous freedom of his dive. Then his curious mind began to wonder about other things. What would happen to a diver if there were a malfunction? Suppose the oxygen supply were suddenly cut off? Could the diver get back to the surface, or at least signal to be hauled up?

Slowly, as though with a force of its own, his hand moved toward the small handle that would cut off his oxygen. He turned it off and instantly blacked out. He knew nothing else until the alert tender of his line, feeling no response, pulled him out. Then came the nightmare! He felt as if he were being approached by an army of men who pounded his ribs, causing terrifying pain. His stomach muscles tied into hard knots; convulsing, he tried to jump out of the boat, fighting the friends who restrained him. The muscles in his throat tightened into round steel cables, then suddenly relaxed and gushed a torrent of blood.

As suddenly as it had begun, the pain ceased and Fleuss became rational. He had cut off oxygen, and without that vital food his entire nervous system had blacked out. In his too quick rise to the surface, the air under pressure in his lungs had expanded and ruptured the lung linings.

He recovered quickly, since there was no irreparable damage, and in a few weeks he was diving again—but wiser after the first near-disaster. About this time it occurred to him that his diving apparatus would be useful in entering flooded mines, or perhaps those filled with noxious gases. A short time later he had the opportunity to offer his invention to Alexander Lambert, a diver who had become a legend in his own time.

Sealab II aquanauts prepare to submerge the habitat off La Jolla, California. During the 45-day project, in 1965, three teams of ten men lived at 205 feet below the sea. (*U.S. Navy*)

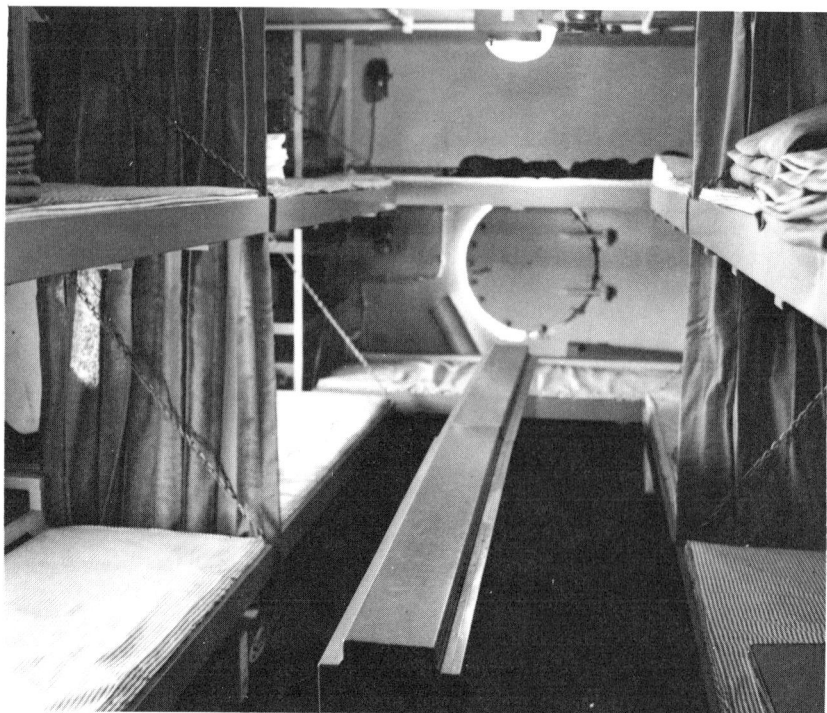

Artist's concept of the Sealab III habitat and support vessel. The support ship will supply electric power and other necessities for the aquanauts living and working at a depth of 600 feet in the San Clemente Island Test Range off southern California. (*U.S. Navy*)

LEFT BOTTOM: The sleeping quarters of the Sealab II habitat provided bunks for ten aquanauts, a large dropleaf table, storage lockers, and an emergency escape hatch. (*U.S. Navy*)

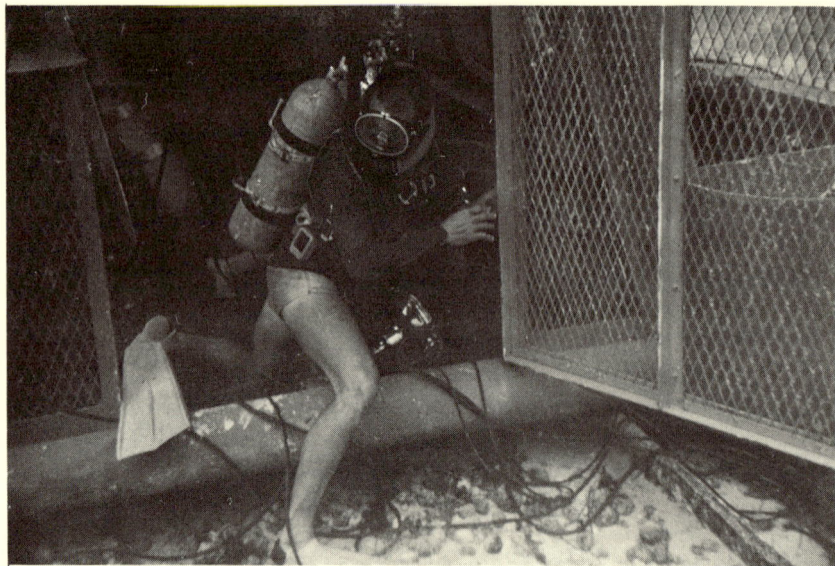

An aquanaut enters the shark cage of Sealab I, off the coast of Bermuda. The Navy's first underwater living experiment, Sealab I, was conducted by four men in eleven days in 1964. (*U.S. Navy*)

The *Star I* submarine lands on a simulated hatch structure during the Sealab I experiment. This test proved the feasibility of using small submarines for rescue operations from larger ones. (*General Dynamics, Electric Boat Division*)

Cutaway of manned underwater station shows five-level living and working quarters under development for the U.S. Navy. (*General Dynamics, Electric Boat Division*)

This personnel transfer chamber is part of a new salvage system being developed that will allow the raising of large sunken objects, such as aircraft and disabled ships. (*U.S. Navy*)

RIGHT BELOW: Electronic instruments such as the "finger probe fish" enable oceanographers to obtain detailed profiles of the ocean floor. (*EG&G International*)

The Perry Cubmarine disembarks at Palomares, Spain, to aid in the under-water search for an unarmed nuclear bomb. The bomb was lost in 2,850 feet of water off the Spanish coast after the collision of two American aircraft. (*U.S. Air Force*)

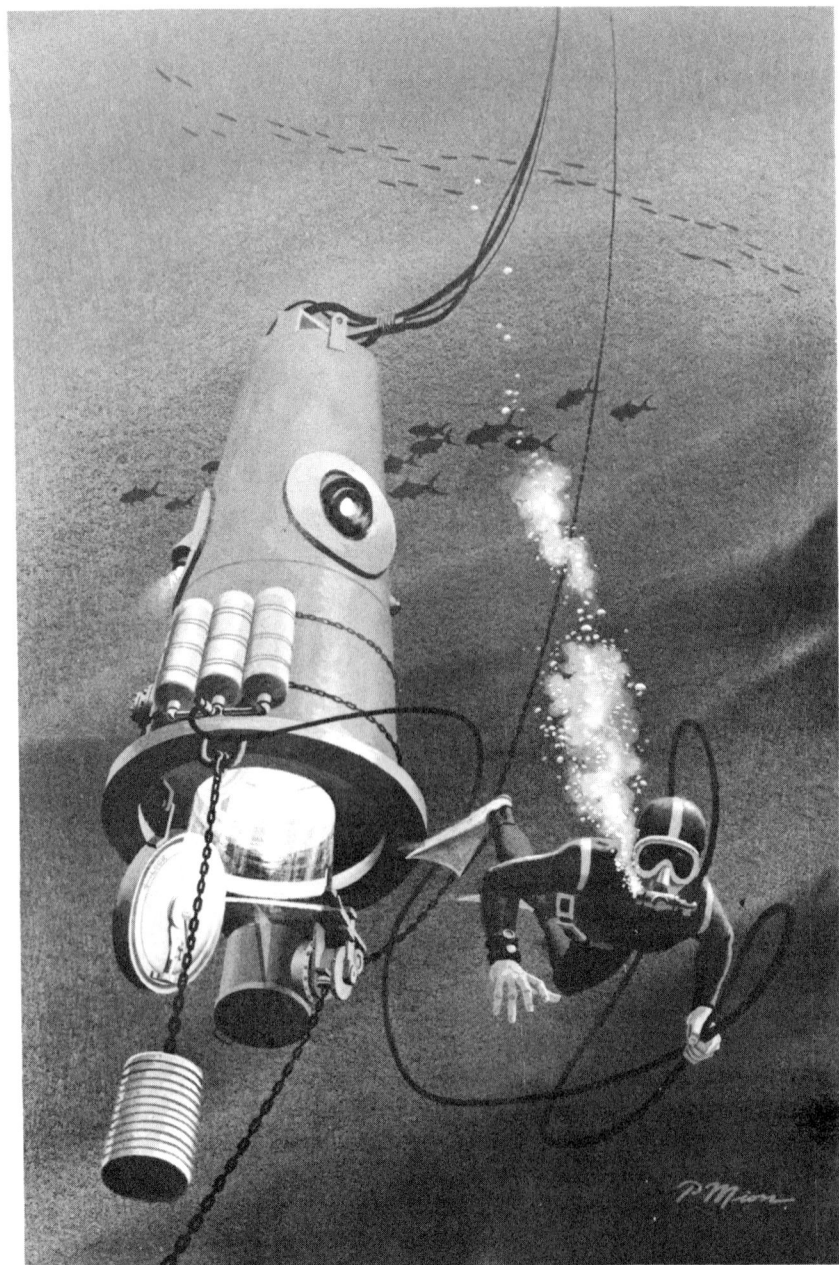

This observation post and way station for divers is on the drawing boards at Union Carbide's Ocean Systems Division.

The submarine *Asherah,* owned by the University of Pennsylvania and built by the Electric Boat Division of General Dynamics, is equipped with an EG&G camera and lights system for archaeological work off Turkey.

France's first underwater habitat, *Conshelf I*. (*Les Requins Associés*)

General Dynamics' nuclear-powered submarine, *Casimir Pulaski,* shows a high degree of maneuverability in a tight 180-degree turn. For indication of the size of the vessel, note the two men in the conning tower. (*General Dynamics, Electric Boat Division*)

Lobster boats and traps in Nova Scotia.

In 1880, a tunnel being excavated under the Severn River in Gloucestershire flooded during a river breakthrough. Work came to a complete standstill, since there was no way engineers could pump the tunnel dry, unless a heavy iron door was closed deep inside. It seemed an impossible job for a diver, who would have to trail air hose and lines. The job required the diver to drop 200 feet down a vertical shaft and then proceed 1,000 feet into the tunnel! Alexander Lambert had been called in and was trying to figure out a way he might accomplish the task.

Hearing about the Severn dilemma, Henry Fleuss offered his new oxygen diving gear to Lambert. Since then science has shown that oxygen can be fatal when breathed more than thirty feet underwater, but at that time neither man knew this. The great Lambert took very little time with his decision after Fleuss explained the use of the equipment.

The inventor helped Lambert strap himself into the equipment, and for what happened afterward there is no scientific explanation. It can only mean that drastic physical variables exist in individuals. By all knowledge, Lambert should never have survived.

The great diver clambered into the black 200-foot pit, tramped 1,000 feet in total darkness along the tunnel's railway, until he came to the open door. The door was jammed ajar, refusing to close. Lambert sank to his knees and groped in the shadowy vault until he discovered the rails that ran over the sill of the door. He grabbed hold of the first rail and ripped it away with his bare hands. The second rail refused to yield, even to his prodigious strength. Lambert turned, trudged his way back to the shaft, climbed up and borrowed a steel bar. Back he went into the water, pried up the second rail, slammed the door, and bolted it securely.

Three years later the Severn tunnel flooded again, and Lambert was asked to go in and close the iron door. He went down again, using Fleuss's gear. This time he nearly died of oxygen poisoning before he found his way out. He recovered, but went back to using a helmet suit for his work.

When, in 1660, the Englishman Robert Boyle wrote his *New Experiments, Physico-Mechanical, touching the spring of air, and its effects*—the book that explained the barometer and contained his famous statement on the relation between pressure and a gas's volume—he had no idea how much he was contributing to the history of diving. Boyle built a compression chamber for the purpose of studying animals in it. One day, in 1670, after he had abruptly lowered the air pressure, he noticed a peculiar phenomenon in one of his subjects. A tiny bubble had appeared in the pupil of a snake's eye. It was a bubble of nitrogen, one of the many that break out of solution into froth when a diver comes from the deeper water into the lesser pressures of shallow water. Ironically, underwater men had to wait two hundred years, and watch many of their number die of the "bends," before a great scientist discovered the meaning of the bubble.

A contemporary of Boyle's built a pressure chamber as a kind of human "heal-all" and called it an "Air Bath." Other quacks picked up the idea and advertised cures for chest deformities, catarrhal deafness, tuberculosis, and prolonged menstrual periods in young women. They clamped their "patients" in pressure tanks as large as waiting rooms, where as long as they were under pressure they obtained relief. The compressed air had an effect on joints, worn lungs, and eustachian systems that could be described as "tightening," but when the patients returned to normal atmosphere, their miseries returned.

In 1866 Dr. Alphonse Gal, the first medical man who became a diver to study man underwater, took Paul Bert, a French diver, to the Aegean and introduced the first diving suits to the Greek sponge divers. The gear they used was the efficient Aerophore invented by Benoist Rouquayrol, whom we discussed earlier in this chapter. Some books also give a Lieutenant August Denayrouze equal credit for the invention, although I found no such name when I searched the actual patent. At any rate, the Aerophore was the best of its kind, and even the revered Jules Verne mentions the name of Rouquayrol in a speech by Captain Nemo to Professor Aronnax.

Bert studied Dr. Gal's comprehensive clinical reports on divers who were crippled or killed in the sea. His resulting and most important discovery was the effect of nitrogen breathed under pressure, which for the first time explained the "bends." The air we breathe consists of approximately four-fifths nitrogen. A diver 33 feet down (one atmosphere) breathes twice as much nitrogen as he does on the surface. At 66 feet he breathes three times as much nitrogen, at 99 feet he breathes four times as much, and so on. This heavy, inert gas does not pass off with the other lighter gas products of respiration, but goes into solution in the blood and tissues, particularly in fat and cartilage. As long as the deep diver stays under heavy pressure he is safe, but when he rises into lighter pressures, nitrogen comes out of solution in the form of tiny bubbles, such as the one Boyle saw in the eye of the snake. These bubbles expand more and more as the diver rises, until they start blocking off capillaries, veins, and arteries. They damage the nervous system and block the spinal cord. The "nitrogen froth," even in its most mild form, causes such intense pain in the joints that the diver stoops grotesquely to relieve it. Hence, the word "bends."

Bert, fortunately, found two answers to the problem of nitrogen froth. One was to have the diver come up very slowly, so that the bubbles could gradually dissolve and the body could pass off the nitrogen while still underwater. This was called "decompression." If, for some reason, the diver had to be brought up quickly, he would be placed in a compressed-air chamber where he would be artificially returned to a deep-water atmosphere until the nitrogen bubbles dissolved. This is "recompression." Bert estimated that the diver could slowly be decompressed at a rate of twenty minutes of decompression for each atmosphere (33 feet).

The next man to carry on the work of Paul Bert was a Scotsman named John Scott Haldane. He was an energetic, deep-eyed man with a granite chin and a huge scraggly moustache. A doctor of medicine, he was originally less concerned with diving than with what foul air did to people in slums, factories, sewers, and mines. As he became interested in diving, another medical man, Dr. A. E. Boycott, joined his team, as did Guybon Damant, a twenty-five-year-old diver from the Isle of Wight, and Andrew Yule Catto, Damant's diving instructor. Haldane took instantly to the quiet, efficient, older fellow-Scotsman, Andrew Catto, and it was only some years later that he discovered that his friend was the brother of Lord Catto, governor of the Bank of England.

The team made many dives during its tenure of activity, all in deep water. Damant made a final dive of 210 feet, a record which lasted for eight years. Exertion tests were made by Andrew Catto in 180 feet of water. Helmet diving equipment was improved. Divers collected their exhalations in a glass vial so Haldane could analyze the gaseous content. After the work of the Haldane team was completed, it was up to science to discover new, safe breathing mixtures, if men were to plumb deeper the secrets of the sea.

In 1919 one of the most prolific inventors of all time, Professor

Elihu Thompson, dropped a note to the U.S. Bureau of Mines suggesting that helium or hydrogen be used in place of nitrogen in the breathing supply of deep-sea divers. He felt that they could go fifty percent deeper.

Helium is odorless, tasteless, and the lightest gas next to hydrogen. Helium is also a nonexplosive gas, while hydrogen is highly volatile. In 1900 helium cost $2,500 per cubic foot. Then it was discovered in mammoth quantities in natural gas wells in Texas, which brought the price down substantially.

The Bureau of Mines joined the Navy in conducting diving experiments in Pittsburgh, Pennsylvania, in 1925. They built compressed-air tanks and began at first to test animals with helium-oxygen mixtures. The animals remained frisky and healthy, so the experiment graduated to the testing of human beings. The humans withstood simulated depths, breathing a mixture of 80 percent helium and 20 percent oxygen. They found that they remained clearheaded in depths where compressed air stole a man's reason. Helium took no effort to breathe, but it did make the subjects feel rather cold. The new gas also had a strange effect on the sounding cavities around the larynx, changing the voice range from baritone to soprano. The tests continued, and in 1937 a young U. S. Navy doctor named A. R. Behnke, along with his colleague, Dr. O. D. Yarbrough, brought a helium diver, clad in a flexible helmet suit, down to a simulated depth of 500 feet, far beyond the tolerance of a man breathing compressed air. Even at that depth the diver felt no unreasoning dizziness (nitrogen narcosis or "rapture of the depths"). He was clearheaded and responded logically (in his soprano voice) to all questions put to him by the doctors.

That same year, on a bleak December day, a young engineer, Max Gene Nohl, had himself lowered into Lake Michigan to a depth of 120 feet. He wore a suit of his own make, but most important of all was the fact that he was breathing a mixture of oxygen and helium from two tanks strapped to his back. In other words, he wore self-contained diving gear.

The next depth record took place ten years later, several thousand miles away from Lake Michigan. In 1948 English Petty Officer Diver First Class Wilfred Bollard, off the deck of HMS *Reclaim* in Loch Fyne, attained a depth of 540 feet. He had to decompress for eight and a half hours.

Nineteen forty-eight was an important year in perfecting the escape hatch for trapped submarine crews. Decompression time was tested and lowered substantially. For example, in a fifteen-minute dive of 180 feet, Haldane called for thirty-five minutes of phased decompression; the U.S. Navy specified thirty-two minutes; and a young Englishman worked out a table for doing it in nine minutes. The experimenter's name was Cyril Rashbass. He also discovered that the deep-water dives of longer duration needed more decompression than the present tables asked for.

While scientists and other dedicated professional divers were developing safer methods for diving, other less altruistic men were utilizing these improvements and turning them into dollars, through underwater salvage. We are not going to spend a lot of time discussing underwater salvage, but a few of the more important jobs are at least worth a mention at this point.

From 1917 to 1924, 25 million dollars' worth of gold was brought up from the *Laurentic* wreck. In charge of operations were G. C. C. Damant (the same man who worked with Haldane) and E. C. Miller.

In 1921 the *Egypt*, with a cargo of six million dollars' worth of gold and ninety-four human beings, was sunk in a fog off Brest. She sank in seventy fathoms of water. The first of her gold was brought up in June 1932, through the efforts of Giovanni Quaglia. It took four more years to bring up the entire six million.

A few years later the mail liner *Niagara* was only thirty miles out of Wangaroa, New Zealand, when she plowed into a German mine and sank in 438 feet of water—with ten tons of gold brick aboard, valued at 12 million dollars. Captain J. P. Williams and James Herd were sent to recover the loot. There was nearly trouble when another German mine became entangled in the salvage ship's anchor line. But underwater blasting and diving continued apace, and in sixteen weeks from the time the salvagers had left port, they had recovered all but six percent of the gold.

These sunken ships were only a few of the many treasures the sea contains. She has swallowed up whole cities, claimed innumerable trade ships. The development of diving technology has led to another facet of underwater activity. This is, at least to my personal taste, the most romantic and exciting one of all: underwater archaeology.

During my career as a diver, I have been privileged to spend some time at this pursuit and fortunate enough to have made some rather important discoveries.

9 · *Underwater Archaeology*

I was up before anyone else, just after the summer sun flooded the early dawn. That early, the day was made up of sounds, not people. Coming from the hills, a high thin voice keened like a distant wind. It was a farmer hawking his fruits and vegetables. Nearby a goat blatted his annoyance with life in general. The sea whispered a sibilant song while it looted a nearby shore.

It was the summer of 1967, and I had returned to the Greek island of Melos, where five years before I had discovered an ancient underwater city. We were here now with a full complement of personnel to film the original story for the Storer Broadcasting Company, as an hour-long special for television release. The sun had just risen, and it would be at least an hour before Fenton McHugh, the producer, Jack Willoughby, our very talented cinematographer, and the rest of the crew would awaken, and I wanted that time to stir for a moment the ashes of recollection before they merged with the new experiences of the day.

I climbed the white, dusty hills to a favorite spot I remembered from years before. It was a cliff from which could be seen an indigo sea, with rocky islands soaring out of its depth. From this lofty perch the sea seemed to have no end. The sun shimmered on its surface, bringing it to life.

I had a strange, nostalgic feeling as I sat there. I was anxious for the day to begin, yet reluctant at the same time. New experiences have a way of dissipating, even disenchanting, the richness of old memory.

As I sat there looking out over the water, a phrase from Gray's "Elegy Written in a Country Churchyard" came to me. It describes the feeling of the Mediterranean better than any other line of poetry. It goes, "Full many a gem of purest ray serene/ The dark, unfathomed caves of ocean bear."

There is no body of water like it in the world. The Mediterranean is a thesaurus of history and stories—stories, bewitching and spellbinding, taken from the pages of a history rich with the lore of the wonderful world beneath the sea. Buried in the beauty of these waters are treasures beyond the ken of present-day values: statues wrought by artists who, perhaps, never knew fame—and statues by those who did; vases, plates, and perfume jars, amphoras filled with wine, and jeweled ornaments. All these artifacts were stored aboard the vessels that plied these waters and sank with the ships when they foundered in a storm.

I remembered that day in 1962 when our small crew of three began to work the ruins of our just-discovered underwater city. After the day's diving we sat with the people from the village of Adomas and discussed methods of search. Underwater archaeology is quite different than that of land. The technique is much more complicated. It was a favorite time of day for all of us and everyone had a story to tell as he clamored for the attention of the group. Ouzo, wine, and beer were poured until they threatened to over-flow the sea itself. Those who weren't talking were singing, or playing a variety of musical instruments.

All became respectfully quiet, however, when an ancient fisher-man rose to speak. I had seen him many times early in the morning on my way to our small diving boat. A venerable resident of the village, thin and brittle, with a face like the subject of an old print, he spoke to the respectful silence of those who really listen.

It was a true story he told us, and the young science of archaeology might have been founded on that tale. As well as I can recollect, this is the way he told it.

"It was early spring in the year 1900. A violent storm was blowing and the blue Mediterranean was whipped into an ominous gray froth. Two Greek sailing galleys were burying their prows deep into the hurling, heaving seas. In the galleys were Greek sea-men and divers from the tiny island of Symē, in the Dodecanese. Not ordinary men, these, for Symē has been famous for thousands of years as the home of the great sponge divers. At the helm of the lead ship stood Captain Demetrios Kondos, a strong, square man, master diver and commander of the galleys. The ships had just completed a generous sponge harvest in Tunisia. The crew had plenty of drachmas in their pockets and were returning home to Symē. A howling, demented northwest wind coming from their port beam pushed them off course, and Captain Kondos sought shelter on the lee side of the island of Antikythēra, near Crete.

"The two galleys anchored under a seventy-five-foot cliff on the far side of the headland, and while waiting for the wind to die, they decided to see if there were any sponges in the water below.

It was Elias Stadiatis who got into his helmet suit and jumped into the misty green water. He slid through the extraordinarily clear water and landed in a land of fantasy. For a moment his mind refused to believe what he saw. Around him in the white sand pranced gigantic white horses; a nude woman stared at him coyly with sightless eyes; a bronze athlete reached toward him with a muscular arm; buried in the sand, a dark hand pointed toward the surface. In a dizzying trance Elias trudged slowly toward the beckoning hand and took hold of it, tugging with all his strength. A heavy arm came up from the sand. He signaled to be pulled up, cradling the arm gently with his own.

"Captain Kondos looked at the arm and called immediately for his dresser. Before his face plate was closed he called for a tape measure, which he clutched in his hand as he descended into the sea. He had been down for a long time when, finally, the awaited 'four tugs' ordered his return to the deck of the galley. The first thing Kondos did when he was able was to jot down several measurements on a scrap of paper; then he looked up at the sky. The wind had all but died, and the galleys sailed at once for home, with the arm resting safely on the bunk of the captain. Back in Symē the crew scattered throughout the island with stories of the strange and wonderful find.

"Captain Kondos called a meeting of the elders of the island.

"The elders, too, had been sponge divers in their time. They all gathered at the house of Mathon Benakis (who was their spokesman), awaiting Captain Kondos. Finally he strode through the door, accompanied by Stadiatis, and carefully laid the huge arm on a table in the center of the room. The elders moved to the table and stood examining the bronze limb. One of them took out a knife and scraped the patina to reveal the bright metal beneath. Another ran his hand along the skillfully crafted long fingers.

"The decision was reached quickly. The arm must go to the museum in Athens, and Captain Kondos, along with Elias Stadiatis, would have the honor of taking it there.

"And so the two Greek sponge divers, dressed in their Sunday best, journeyed to Athens, where their prize was received with enthusiasm. Quickly they were commissioned to return to the site and take up the difficult and often dangerous task of salvage. It was decided that their remarkable find was most probably a Roman plunder ship returning to harbor with stolen Greek sculpture.

"Professor George Karo, who inspected the artifacts as they arrived in Athens, was definite in his praise of the divers. 'These illiterate fishermen, totally ignorant of archaeological techniques, treated the finds with quite remarkable care and delicacy. I was amazed at the small amount of recent damage. Not only had the sculptures been handled with evident gentleness, but even pottery and glass vases had been brought up intact.'

"Included in the many finds was a truly fine bronze of a naked, noble youth, with gemstones set in his eyes to delineate the pupils. He dates, perhaps, from the fourth or fifth century, although scholars differ as to whom he represents. Some say Paris, others say Hermes or Perseus, and so he is called 'The Athlete.' "

The old man stopped talking, and for a while the only sound was that of the waves gently spanking the sterns of the fishing boats anchored just offshore. The men nodded gravely, the women smiled fondly, I was entranced by the tale and let my thoughts drift farther.

The Antikythēra expedition spurred great hopes for underwater archaeology, but very few came forth to carry on this somewhat dangerous and exacting science. Archaic equipment, lack of decompression knowledge, and dearth of sponsorship held back the steps of progress.

Yet gradually more finds were made. In 1907 another Greek sponge diver discovered a second Roman plunder ship, full of Greek sculpture. He was about a hundred and thirty feet down, scanning a somewhat barren bottom three miles off Mehdia, Tunisia, when he came upon rows of long cylindrical objects covered with shellfish and algae. His first thought was that he had found a cask or cannon. He looked closer and, after scraping off some of the matted algae, found a marble column. Close by, many other objects protruded from the sand, and he found himself in a treasure house of graceful bronze figures and lovely stone carvings. Back on the surface he described volubly what he had seen, and soon that small spot on the Mediterranean rained Greek divers. They picked up everything they could carry and disappeared into the nearby ports.

Shortly after, Alfred Merlin, Director of Antiquities of Tunisia, discovered that genuine Greek art objects were being sold in the Arab marketplace. A quick investigation revealed that the Greeks were happily recovering and selling what the Romans had taken from them two thousand years before. These were French waters, however, and the jubilant harvest was over.

Solomon Reinach, an influential classical antiquarian, convinced

American millionaire James Hazen Hyde and two Frenchmen, Duc de Loubat and Edouard de Billy, to back a professional expedition to return to the underwater site of the Roman galleon.

Elated, Reinach saw the first discoveries to reach the surface and declared: "Nothing comparable has come to light since Pompeii and Herculaneum." The excavation continued for five years and yielded enough *objets d'art* to grace five rooms at the Alaoui Museum at Bardo, Tunis.

Forty years later the Mehdia wreck served another important purpose. It became the first underwater school of archaeology for scuba divers, when Cousteau, Tailliez, and Dumus relocated her while searching from their ship, the *Elie Monnier*.

So that we have a complete picture of underwater archaeology, let me describe a scene as it might have happened some two thousand years ago: It is the year 50 B.C. and the early morning Greek shore hears the creak of many oars as the galley sails south along the regular trade route. Her upper decks are piled high with amphoras filled with grain, oil, wine, water, and perfume. As so often happens, the small ship looks dangerously overloaded. She proceeds slowly on course as dark, angry clouds begin to build up on the horizon. The captain looks up, worried, and urges greater speed from the weary oarsmen. Then suddenly the storm is upon them and the wind, shoving and prodding, pushes them off course. A rending sound is heard, and the clumsy square sail is ripped off her mast. The oarsmen, fearful of their lives, fight to keep the ship from broaching, but the power of the storm is much greater than their puny human strength. The heaving sea tosses them high, and some of the amphoras are flung off into the water. The bow goes down, burying deep in the waves. The stern rocks high, and when it comes down the tortured, shrieking sound of shattering wood is heard. The galley has come down on rocks just below the surface, and now she takes on water.

She is a wounded thing incapable of controlled movement. Again she rises and crashes down on the rocks. She is stove through her lead plating and Aleppo-pine sheathing. The bare Lebanon-cedar ribs show through like raw bones through a fatal wound. The life goes out of her and, shuddering, she begins to sink. The men leap clear, cling to exposed rocks, but the waves find them and drag them loose. The shattered wreck sinks almost upright and drifts down into the deep water with the dreamy floating motion of a falling leaf. As she drifts to the bottom she strikes ledges that gouge still deeper wounds into her hulk. Then she touches the silent bottom, and a cloud of soft, gray, sandy mud billows up and, for a moment, hides her final death agony. After a time the cloud settles. Slowly the animals of the sea creep forward and claim the ship. An eel settles in a broken amphora; an octopus nestles under a plank; sea fans send forth their tender little roots; teredo worms take the first destroying nips; a large mero wanders by and takes over a dark hole as his new home; a large shark pauses curiously for a moment, then glides on.

And for a while time seems to stand still.

But time is the patient artist who changes the ship's appearance completely. Years later, her mast and wooden hull have disappeared. Light-colored mounds have erased her form. She has become a world of colorful fish. Red sea fans, full grown, wave gently, like historic flags. Yellow sponges and green algae have changed the ship from a manmade form to a form with the natural look of the underwater world.

There are some things, however, which have not changed. The lead and bronze hardware has resisted all corrosion. Nothing has been able to eat the earthenware dishes or the amphoras, made of eternal fired clay.

This was the ship's history for two thousand years, until a modern diving archaeologist discovered her and brought her story to life.

Just one hundred years ago the Japanese invented their own method of recovering art objects—at least their own art objects. An open boat loaded with the emperor's favorite vases went down in the Sea of Japan. His nibs dispatched a boatload of divers, with orders to recover his little baubles. The naked divers, however, found the depth beyond their capabilities. Naturally, they didn't dare offend the Son of Heaven, and so they did some quick and heavy thinking on the spot. The solution was a unique one. They captured a number of live octopuses, tied lines to them, and lowered them into the wreck. These eight-limbed creatures are very fond of confined spaces, so they quickly slid into the drowned jars. The Japanese brought up the lines—octopuses, jars, and all.

Since the invention of modern diving equipment—especially the compressed-air tank, demand regulator, fins, and face mask—underwater archaeology has definitely been stepped up. Treasure diving also has amassed large fortunes for some lucky individuals. Diving changed from a heavy, plodding, limited activity to a free-moving, mobile, "soaring like a bird" science.

And that was why we were here on the island of Melos. We wanted to utilize this diving mobility and write a small page in the history of archaeology.

It was time, to start the day. I scrambled down the hill to wake the crew. Within minutes all was a bustle of activity. Johnny, who was assigned to us by the Greek navy, began to fill our bottles (compressed-air tanks). Fenton McHugh, Jack Willoughby, and I had a conference at which I diagrammed how we would accomplish our first days of underwater shooting. Marco Rosales, my assistant diver on this job and an accomplished musician, provided a background of soft guitar music. As a matter of fact, every time there was a pause in the diving activity, Marco's guitar magically appeared.

Breakfast, prepared by two motherly Greek ladies, was eaten outdoors. Directly afterward, while everyone was still assembled, I explained the most important "do's and don'ts" of underwater archaeology:

"Very soon we are going to be at the site of the ruins of an ancient underwater city. It is most important that no one remove any artifact that he may find. There is a very good reason for this. Once an artifact is removed from where it is originally discovered, not only is the artifact out of context but the site itself may be lost forever. It is what lies *beneath* the amphora you find that is important, because that is where the bulk of the find may be buried by the sand.

"Normally we would construct a grid over the entire site and number each square. Then pictures would be taken and sketches made. This allows for a safe, careful, methodical search.

"We would use an airlift (which looks and acts like an underwater vacuum cleaner) to suck away the sand, revealing what lies below. Baskets would be lowered from the boat above, and all discoveries carefully placed within would be taken to the surface. Any artifacts taken from a numbered grid (square) would be meticulously recorded.

"Fortunately, except for the airlift, most of this work has already been done. We are merely going to recreate the discovery of the underwater city exactly as it happened five years ago. At that time we had only a three-man crew and even less equipment.

"The most important thing for you to remember is if you find an object, such as an amphora, please leave it where it is, and come and get me. I'll decide whether or not it can be removed."

Using two local fishing boats, we headed for the site of the underwater ruins some three miles away. On our last visit we had completed our undersea search by bringing up a most exciting find. I had discovered what I at first took to be a milling bowl almost entirely buried in the sand. As I slowly uncovered the object, I found that it was about three feet high and two and a half feet in diameter, with a hollowed-out bowl perhaps a foot deep. It was much too large to be a milling bowl, and it was only when we raised it and brought it on deck that an archaeologist, Dr. Vaos, identified the object. It was, he said, a sacrificial bowl which most probably dated back to neolithic times. Our discovery was approximately six thousand years old.

At that time we knew every wall, every stone of the excavation. I wondered now, as we slowly chugged our way back to the location, if I would be able to find the site again without a time-wasting search. I stood on the bow and stared at the water. It was still dark blue, still too deep to be over the ruins. Then it began to change color. The blue mixed with green. The bottom was sloping up. We were getting closer. Scanning the shore line, I spotted familiar landmarks. Near the shore were the remains of a very old windmill. To the right was a line of rocks forming a small breakwater. To the left was a small peninsula. The three lines extending from these landmarks met somewhere offshore. This was my fix. I signaled for a stop and we threw off the anchor. The ruins of the city were supposed to be just underneath. Normally, we would have been able to see even from the surface, but a recent storm had stirred up the bottom.

The rest of the crew thoughtfully held back while I donned mask, snorkel, and fins. They were waiting for me to have the pleasure of the first look.

I jumped over the side and headed down. Although the visibility was not as good as it should have been, I spotted the walls immediately. Everything was exactly as I had left it years before. Surfacing, I signaled for the men to look for themselves. It didn't take long. There was a quick flurry of splashes and everyone was in the water, looking for all the world like a school of dolphin frolicking at Marineland.

Reluctant to leave, yet knowing that we must get on with the always complicated business of underwater filming, I gave myself the luxury of fifteen more minutes of exploration.

A few thousand years ago this city had been built on a tranquil hillside that sloped gently to the shores of the sea. During that period in the world's history, two major earthquakes occurred each hundred years. Violent quakes like these caused the fabulous palace at Knossos on the island of Crete to be rebuilt many times. The island of Melos lies deep in the Cyclades, midway between Crete and the mainland of Greece. Moreover, it possesses one of the best natural harbors of any of the Cycladic islands. During the Minoan period (3000 B.C. to 1250 B.C.) the island of Melos must have been of consummate importance. Sailing vessels stopped here on their way to and from Egypt. Cities flourished.

Then an earthquake struck Melos, violent and demolishing! Entire sections of cliff sheared off, tumbling people, buildings, and domestic animals into the water. Large new areas of land forged miraculously out into the water, while other land masses sank.

It was during one of these violent quakes that our city slipped beneath the cool clear water. Storms, too many to count, shifted the sand until the city looked as it did now, a complex of walls standing about three feet above the sandy bottom. But below the sand were the remains of two-story dwellings and other more important buildings—villas, perhaps, and bakeries and warehouses. Emborium had been a seaside settlement of perhaps two or three hundred people.

But we had a movie to film. The water was not clear enough for good photography; suspended matter needed another day to settle back down to the bottom. I knew of another location across the harbor where the water should be clear. Actually, this is where we had begun our original search years before. Marco Rosales helped me herd the others out of the water and into the boats.

Forty-five minutes later we were at the other location. A rocky cliff soared out of the water a thousand feet above our heads. The cliff was pockmarked with caves, both below and just above the surface of the water. These caves were not a natural phenomenon. They were made by the early Christians as hiding places from the Romans.

We got Jack Willoughby ready for the water first. He had had very little experience with underwater photography in the open sea, but he mastered the inherent difficulties faster than any man I have ever worked with. As soon as he was ready, he went over the side, and we handed him the underwater motion picture camera. It was the one I have been using with excellent results for some years, a 16-mm. Bolex made by the Paillard Corporation.

Then began the arduous task of filming an underwater adventure in the realistic environment of the sea itself. There were no safe, calm Hollywood tanks for the actors to perform in. The camera-man, too, was *in* the water, not shooting through a glass port while sitting in a dry room. When the wind came up in the after-noon we were buffeted by the surge, our legs and arms bruised from hitting the rocks. But the footage we shot reflected the realistic situations incurred when doing a film on underwater archaeology. As usually happens in this kind of situation, everyone bore the marks of battle with no small pride. The very difficulty of the situation welded the men together as a well-functioning team.

On the third day we decided to explore a special cave located about eight feet above the water. We shrugged out of our diving equipment and left it at the base of the cliff. A camera was handed up to Willoughby, who wore a battery pack belt for portable lights, which were also handed up. Willoughby, his assist-ant camerman, Marco, and I huddled at the entrance of the cave for a quick conference. We decided to shoot the scene "off the cuff." We would let the camera discover what secrets the cave held at the same time we discovered them.

We entered and I knew immediately where we were!

"It's a crypt, Jack," I whispered, for no good reason. There was no one around who cared whether I whispered or shouted. "Mostly probably early Christian. Let's dig a little and see what we can find."

Jack sprayed the light around the dim interior. There were five burial places. Marco and I knelt down alongside one of them and began to dig into the sand with our bare hands. First we uncovered some personal effects of the deceased—plates, vases, and similar objects. Then we came to the bones, yellow and hard with age. We photographed everything, then laid the bones and artifacts back to rest. Carefully we smoothed the sand over the grave and restored it to its original look of rather wild, timeless peace. Stormy waves had scattered strands of seaweed over the floor of the crypt and the warm summer air had dried them. These, too, we replaced on the graves.

There were writings on the rough walls and crudely drawn pictures. One picture depicted an angel in a long flowing dress. We recorded all on motion picture film, then turned to leave. Once more we felt the sun on our backs and blinked in the white-hot brightness of the day.

Greek summers are flawless as far as weather is concerned. Although they are hot and humid it does not rain. We were in the water filming all day and every day for two weeks, and then the underwater phase was over. Another week of surface photography, and we were ready to pack up and leave. You pack up equipment and clothes, but not the memories. For a while they stay with you, clear and cherished. Later, when you are with friends back home, you take them out and proudly show them like a favorite picture album.

Underwater archaeology is vastly different from archaeology on land. An excavation on land is a relatively simple chore, and it reveals stratification that is rather easy to read. Excavating underwater is a much more difficult, extremely frustrating experience. First of all, besides the divers, scientists, and considerable support personnel, it requires a good deal of heavy equipment. As you begin to dig with a dredge or an airlift, for a while the situation may look encouraging. You are obviously over the wreck of an ancient ship, because as you struggle to keep the hose of the airlift pointed in the right direction, you are rapidly uncovering rows of amphoras stacked one on top of another. When the hole gets to a certain size, however, you find that the sand is sliding back in as fast as you are taking it out. You may not even realize this, because by now the water is so murky with stirred-up bottom sediment that you can see absolutely nothing around you. You might also be unknowingly digging yourself into a hole and be in danger of being buried alive.

In 1871 an ardent but amateur archaeologist named Heinrich Schliemann dug a great trench at Hissarlik, the theoretical site of Troy, in Asia Minor. The trench cut through the stratification of 3,000 years and revealed the remains of *nine* cities, one on top of the other.

This trench did not collapse, as it most certainly would have done underwater, but contained itself with the help of simple shoring.

It is also very possible for the entire cargo of a wreck to be firmly embedded in rock-hard concretions. Under these circumstances large chunks must be chipped free and raised to the surface. Under controlled conditions further chipping will take place and the essential parts will be ultimately fitted together. These and other finds, however, are well worth the effort, and their historical contributions are undeniable.

The chance of finding treasure—coins, jewels, gold or silver bars—on a wreck site is not as meager as one might suppose. Countless pieces of eight, gold doubloons, gold ingots, silver bars, and jewels have been found in accessible waters as near the United States as the Florida coast and the Bahamas. There are a number of books, such as John Potter's *Treasure Diver's Guide*, which give possible location sites for treasure diving. Potter's book is especially well researched and cites authentic locations.

There are two basic methods of search, visual and electronic. We'll take up the subject of visual methods first.

The boat is moored over a likely spot and a heavy weight or anchor is dropped over the side. The divers swim down with a predetermined length of rope knotted at regular intervals. The rope is attached to the weight, and the divers swim in concentric circles, stopping at the completion of each circle to place a marker before moving down to the next knot. The search of the area is methodical and complete.

Another way to vector an area is by having a group of divers swim *in line*, all searching the bottom.

A simple, inexpensive sea sled towed at a speed of about four knots will cover a large area quickly.

A glass-bottomed boat, or simply a bucket with a glass bottom, will allow you to make some kind of search without even having to get wet. This is not very effective, however.

A slow-flying small plane cruising at low altitudes can provide an effective method of spotting a wreck, providing, of course, that the water is clear.

A few years ago Teddy Tucker searched the coast of Bermuda while dangling from a balloon.

One of the newest and best methods of search with electronic equipment is the one developed by Dr. Harold Edgerton of EG&G, Inc. This very sophisticated method is accomplished by trailing a hydrophone and transducer, which results in a seismic profiling of the ocean bottom. Graphs or pictures of the bottom are provided for leisurely scrutiny and interpretation.

Other electronic methods include the old mine detector, which has very limited application, or the more efficient magnetometer, which is secured underneath the search boat but whose signals are read from the surface.

Hand-carried induction detectors are very efficient, when in the known area of a wreck site. An excellent lightweight model of this type is the UML-20 made by the Bludworth Marine Division of States Electronics Corporation in New York City.

Now before you pack up the old kit bag and head for the briny deep, it might be well to conclude your crash course on underwater archaeology with some information as to what happens to organic and inorganic materials when immersed in salt water:

WOODS: If exposed will be attacked by teredo worms and fungi and soon destroyed. Will survive if buried by sand or mud.

HEMP OR SISAL LINES: Will survive only if quickly buried, and then only in fragile condition.

TARRED LINES: Will survive and retain some strength even if not buried.

BONE: Hard, less porous bones will survive in good condition.

IVORY: Usually survives.

LEATHER: Can survive only if buried quickly.

IRON: Corrodes deeply, and unless treated will flake until completely destroyed.

STEEL: Depends on quality. Can survive with little damage.

BRASS: In thick forms, such as cannons, will survive under the green patina.

SILVER: Will revert to carbon dust unless touching another metal giving cathodic protection (higher electrode potential).

COPPER: About the same as brass.

GOLD: Remains in good condition.

LEAD: Remains in good condition.

PEWTER: Variable. Some will be found in good condition and some in bad.

PORCELAIN: Survives very well.

TERRA COTTA: Survives well.

GLASS: Survives well, except for some surface erosion, except *green bottle glass*, which does not usually survive in good condition.

COAL: Survives well.

MARBLE: Usually survives well with little erosion, but softer marbles can be affected by a particular kind of sea worm.

GEMS: Most gems will survive very well.

PEARLS: Do not survive well. The form survives but the luster will be dull.

10 · Underwater Photography

In 1892, when Louis Boutan, lecture-master of the Paris Faculty of Science, first announced the idea of taking pictures underwater, the announcement was greeted with silent, Gallic shrugs by his colleagues and loud derisive catcalls by those outside the hallowed halls.

Nevertheless, Boutan went to work and built an underwater case for a fixed-focus camera. The copper, waterproof box had one external control that operated the shutter. It also had three glass ports, one for the lens and two for the viewer. Already an experienced diver, Boutan pressurized the case, so that it would not collapse underwater, with a simple but ingenious gadget. He ran a tube from the box to an inflated rubber balloon. When the water pressure collapsed the balloon, it forced more air into the camera case. One disadvantage was the time it took for a proper exposure. The diver-cameraman would turn the lever that opened the shutter. He would then have to wait between ten and thirty minutes before he could turn the lever the other way, closing the shutter.

Boutan made his first photographic dives in 1893, and three cameras later he made his first successful picture. Among other valuable facts, he discovered that an underwater lens received a refracted image, in which the subjects appeared one fourth larger and nearer than they actually were. This is also true of what the diver sees behind his glass mask and is due to the different speeds of light in water as compared to air. Boutan's third camera had an astigmatic lens that could be focused before diving. It took photographs seven by nine and a half inches. His successful pictures began to evoke interest in the public, though they could see very little use for taking pictures in the wastelands of underwater. A French newspaper published a cartoon that suggested a reason for Boutan's avid interest in underwater photography. The caustic drawing depicted a rather mad-looking professor lying on the shallow bottom of a beach photographing the luscious limbs of bathing beauties. In a clever, amused riposte, Boutan blasted the paper by taking a picture of three sturdy sailors in long striped drawers standing waist-deep in water. He sent the picture to the newspaper, which good-naturedly printed it and allowed all France to laugh.

Boutan was working on a book (published in 1900) sometime during the summer of 1899, when he had a visitor—one Monsieur Deloncle, president of a large and prosperous optical manufacturing company. His engineers had built a giant siderostat telescope for the 1900 Paris Exposition. It was Deloncle's idea to pair pictures of the planets in outer space with photographs from the ocean depth. He could not have arrived at a more opportune time, since Boutan needed money to carry on experiments with illuminated underwater pictures.

However, Boutan hesitated before he accepted the commission. He explained to Deloncle that a kind of blue filter existed underwater which could not be penetrated by natural light, but artificial light removed that filter and enabled the underwater scenery to be viewed in all its lovely vivid colors. But the artificial lights existed in theory only. Deloncle agreed to furnish the funds for their development providing that the photographs were ready for projection at the Optical Palace at the Exposition.

Boutan built two storage batteries with an output of twenty-five amperes per hour. They furnished power for two submerged arc lamps, each with a gap of twelve inches to be spanned by a bright arc flame. It was estimated that they would burn for half an hour on a battery load, and they had been tested to a pressure of 330 feet.

The gear was ready for a practical test by the end of August. Boutan decided to make the test at night, and he and his helpers anchored their boat 300 feet from shore. There was no moon, and the sea, though calm, was pitch black. The apparatus was lowered twenty feet into the water below. Then came the moment to test the electrical circuit and take the first underwater picture ever taken with artificial light! Boutan pulled the switch and everyone on board gasped with the pure enchantment of the scene. The sea glowed jewel green, and in the clear, misty depths fished scampered like frightened butterflies. Yellow gorgonians danced on the crest of coral mounds. So enthralled was Boutan with the scene, that he very nearly forgot to trip the switch that operated the shutter.

The first photograph showed a rather spectral image of a sea fan growing atop a piece of coral. Boutan decided to test again at fifty meters (165 feet) and make some changes. A metal framework was quickly built, with the camera mounted in the center and an arc lamp on either side, angled in toward a sign that read "Photographie Sous Marine." Boutan's crew took the boat out somewhat farther from shore this time and lowered the heavy gear to 165 feet. The arcs were ignited and the plate exposed. The result was a sharp, clear picture that astonished and delighted Monsieur Deloncle.

Boutan's work inspired one quick follower, Étienne Peau, a semi-invalid and son of a shipbuilder from Le Havre. He built a round steel camera case, and since he was already an experienced naturalist, he began taking pictures of myriad sea animals. In 1905 he began to study the eyes of the Maya crab in the hopes that he would learn more about underwater optics. The eye of the crab has a kind of conical lens shape, which Peau thought eliminated incident light in shallow water. So he fashioned a lens shade after filling it with filtered water. This had the effect of extending the camera eye almost a foot farther out into the water, thus pushing away the suspended particles nearest the lens, which normally block the view. He liked to take round pictures, which, he felt, had a "peculiar charm, like a telescope peeking at a secret place."

Following Étienne Peau, an electrical engineer, H. Hartman, patented a huge 1700-pound depth camera in England in 1913. This camera was remotely controlled and took clear pictures at a depth of seventy-five feet with a depth of field of twenty-five to sixty feet.

About 1914 a man came along who was the answer to the publicists' dream. He was a cross between P. T. Barnum and C. B. DeMille, with a little bit of Jack London thrown in. His name was John Ernest Williamson, and he had been born in Liverpool of a Scottish seafaring father who brought his family to the United States and started a shipfitting business in Norfolk, Virginia. Young John studied art and became a reporter, photographer, and cartoonist for a Virginia newspaper. A marvelously articulate man, with a great persuasive gift for poetic word images, he became the world's first underwater motion picture photographer and producer. Extending from the bottom of his boat he built a long, flexible tube, at the end of which was a larger chamber from which a cameraman operated through glass ports. Williamson called this the Photosphere and fitted it with submersible mercury vapor floodlights.

By February 1914 he had talked his way into complete sponsorship of a motion picture, and so he sailed to the clear waters of Nassau. One of the scenes he had promised for the film was a heroic "fight to the finish between man and shark." He hired two West Indian divers noted for their fearlessness, lowered the carcass of a dead horse to the bottom, and prepared to film the "never before seen" action through his camera ports. Well, John saw the action all right, all of it, right to the spectacular killing of a shark at the end . . . but all of it just *outside* camera range! There was no way the frantic crew could turn the Photosphere just a little to the left so that they could capture the action on film.

The West Indian divers surfaced, demanded their fee, and after being paid, departed forever. Now this is where the Jack London part of Williamson's character comes in. He was so resolved to get the entire shark sequence on film, that he decided to do it himself. He covered himself with shark oil, which according to the West Indian shark fighting tradition, protected him from attack. After applying it, his crew had to admit the antiattack theory must be true, since the stench was so overpowering that nothing would dare get close enough to attack. By this time there were some twelve sharks cruising in exactly the right position for the camera. John plunged in, and being a natural athlete and a magnificent swimmer, actually accomplished the feat. He killed a shark before the eyes of the camera and of the thrilled audiences who saw the epic, which was entitled *The Williamson Submarine Expedition*. He produced three other adventure-diving sagas, *The Submarine Eye, Girl of the Sea*, and his grand opus, *Twenty Thousand Leagues Under the Sea*.

In 1917 an ichthyologist named Dr. W. H. Longley led a return to Boutan's practice of having both man and camera underwater. He built a heavy brass camera box fitted with exterior controls for focus, speed, and shutter, and began taking excellent 4 by 5 photographs on the brightly lighted floor of the Caribbean.

A giant step forward in camera equipment occurred in 1928, with the development of Commander E. R. Fenimore Johnson's motion picture camera called the Fenjohn.

During the early 1930's a French naval officer, Commandant Yves le Prieur, devised 16-mm. movie units and 35-mm. still cameras to use with his independent compressed-air lung. After that marine photography soared like a bird. Captain Jacques Cousteau produced the glorious, prize-winning *Silent World*. Dimitri Rebikoff, in the early 1950's, built his famous Torpilles Sous-Marines, which contained camera, electronic flash, and electric propulsion to carry the photographer.

Basically, underwater photography utilizes two systems. One system is remote-controlled, with the camera and lights secured to a small submarine or assembled on a specially designed frame. This system is used for deep-water photography in which the camera is activated by a timing device. This is random photography, and as in opening a grab bag, one never knows what one has until the roll is developed.

In the other type of underwater photography, the camera is held by the diver. There is a need for a superior type of diver-held camera with attached light source. The camera, like the lights, should be small and lightweight. The ideal strobe-light unit should be strong enough to photograph subjects in true color from a distance of at least twenty-five feet (although most underwater photography is done in close-up, since it uses a wide-angle lens). The strobe should have the ability to recharge quickly and silently with a battery strong enough to energize several hundred flashes without replacing.

Although the flash unit is a bit heavy, Dr. Harold Edgerton of EG&G International has developed an excellent unit that comes close to the ideal specifications. The unit uses 510-volt batteries, which are somewhat expensive ($15) and not commonly available, so the photographer has to carry an extra supply. The flash capacitor has 50 watt-seconds of energy storage, and a flash reflector combination, which produces a broad beam of light matching the field of a 35-mm. lens. The output is 1,500 beam candlepower-seconds. Underwater at three feet, subject to lens distance, the guide factor would be about 30, so the photographer would shoot Kodachrome 11 film at an f-stop between f/8 and f/11. The unit is four inches in diameter and one foot long, with a metal arm that attaches to the camera. It is very well adapted to use with the Nikonos camera.

For the recreational diver, a good dependable flash unit, which is adaptable to both the Nikonos and the Calypso 33-mm. cameras, is the Aquaflash. It sells for about $65.

Among the 16-mm. motion picture cameras, I prefer the Bolex in its specially designed underwater housing. With it I have accomplished complete motion picture productions.

Any amateur diver, however, as soon as he acquires just an average proficiency in the water, can take exciting, prize-winning underwater photographs. Following are a few words of advice that hopefully will help you record on film your next underwater holiday.

First of all, select the camera from which you can expect to get the best results. The two most popular cameras are the above-mentioned Calypso and the Nikonos. They both offer a more-than-adequate range of shutter speeds and possess excellent lenses. Further, they are both small, simple to operate, and work very well with the flash attachment.

Next, be certain to choose the right film for the job. If the water is rather dark and a bit murky, use Tri-X. In clear water Plus-X, which has an ASA rating of 160, will provide you with clear, crisp negatives. If you are shooting color in clear water with brilliant sunlight, you can use Kodacolor and produce truly beautiful color prints. For color slides, it might be better to stick to High Speed Ektachrome, which has an ASA of 160 and offers more latitude for slight mistakes in exposure.

Use an underwater light meter, and believe the light values you read there. Do not trust your eyes. They will too often tell you that there is less light than really exists, which can only lead to overexposure and a ruined picture. It is generally better to shoot at faster shutter speeds and lower f-stops to avoid the possibility of blurred pictures, although the density of the water will help stabilize your camera hand. If there is surge in the water, find a way to stabilize your position by kneeling or sitting on the bottom. Another good tip is to take off your fins and hand them to your diving buddy, and thus avoid stirring up the bottom. Try, also, to get as close to your subject as possible.

The best natural light is found in shallow water; in depths of twenty-five feet or less you will accomplish your best color photography. Colors, especially red, fall off very rapidly underwater, and without a light source, you can only get an all "blue" picture in deeper waters. This does not mean that beautiful, vivid colors do not exist in deeper water. They are there in all their glory, but to be revealed, they must be illuminated with an artificial light source.

In most tropical waters, such as the Caribbean, the Florida Keys, and the Bahamas, the bottom is composed of light-colored sand manufactured by time from pulverized coral. Especially in shallow water, this sand reflects the light from the surface and furnishes excellent "fill" light.

For your "long" shots (eight feet or more) it is a good idea to shoot your subject from slightly above or slightly below. Avoid "straight on" shots. Even in what appears to you to be perfectly clear water, there are suspended particles. These particles are reflecting light from the surface and will reflect that light directly into your lens if you shoot straight on. The result will be a milky, hazy picture.

There is one final assurance that I can give you as an amateur underwater photographer. There is a very complete and wide price range for both still and motion picture cameras. You will find them advertised in any recent issue of *Skin Diver* magazine. There is a camera to fit everyone's budget.

The *deep* ocean is a totally dark world. Below 1,000 feet there is absolutely no penetration of sunlight. It is essential that man explore this area, for much of the world's mineral wealth is at these depths. We still have not developed acoustical devices that can give us a clear picture of the topography of the bottom (although such devices are reaching a stage of high development) or of the flora and fauna that survive without sunlight.

There was a time in history when men made heroic sea journeys— men like Leif Ericson, Christopher Columbus, Vasco da Gama. They were truly the courageous explorers of their time, and their journeys helped mankind and the world to develop.

Man today is still making voyages of discovery, both in outer space and in the economically more vital "inner space." The great sea journeys now are made underwater to ever greater depths.

Deep-sea cameras are the "eyes" of these underwater sea journeys. Probably the greatest developer of deep-sea camera and light systems is EG&G (Edgerton, Germeshausen & Grier), located just outside of Boston.

This firm's cameras, camera light sources, and incandescent (steady-state) lights have been mounted aboard such famous submersibles as Jacques Cousteau's diving saucer, the *Souscoupe, Trieste I* and *Trieste II, Asherah, Alvin, Star II,* and the Perry Cubmobile.

The camera is a 35-mm. still camera capable of taking 2,300 exposures with a 35-mm. f/4.5 Hopkins with an adjustable focus from three feet to infinity, and a shutter speed of 1/10 to 1/200 second. It is designed to withstand a pressure of 12,000 psi (25,000-foot depth), and while it photographs the subject, it also records the time, date, and frame number reference.

The flash is xenon discharge with a flashtube input of 250 watt-seconds and a flashtube output of 12,000 lumen seconds over input voltage range. It has a flash repetition rate of three seconds over input voltage range.

EG&G has also developed a new multilayer film for scientific photography that may become feasible for use by the recreational diver. For the moment it is called XR film, and it has an exposure response range of 1 million to 1. This would make it almost impossible to shoot a badly exposed picture. I tested the film in diving locations throughout Europe in 1968.

Another great potential of this photo system was illustrated recently when the Perry Cubmarine made a photo survey of AT&T's transatlantic cable. The photographs clearly showed potential trouble spots, such as cable suspension, snags, and damage done by fishing trawlers.

11 · Marine Engineering—Vision and Reality

"The sea drowns out humanity and time: It has no sympathy with either, for it belongs to eternity, and of that it sings its monotonous song forever."

—*Oliver Wendell Holmes*

Mr. Holmes was a wise and talented man, but he had no practical concept of the sea's relationship to man. His words, unfortunately, did influence and reflect the thinking of his time.

For too many years the sea has been regarded as a place to take a refreshing dip on a hot summer day, a means of travel (with apprehension) from one place to another, a means of transporting commodities (with high insurance rates) from one country to another, and a surface to be skimmed over with great daring, cradled in the cockpit of a speedboat.

For centuries we tolerated the ocean waters, admired their beauty, feared their awesome powers, spewed pollutants into their depths, and cautioned our children not to go too near them.

Then came the first rumblings, the first cautious warnings, the first timid alarms. In the 1950's the Paley Commission compiled and published a conservative but upsetting report entitled *Resources for Freedom*, which contained frail warnings of a widening breach between mineral supply and demand.

New words appeared in newspaper and magazine stories containing dire predictions: overpopulation, imbalance, population explosion. In twenty years the rapidly growing population of the world would eat itself right out of existence!

We began to look to the sea as a friend who could save us, rather than as a devourer of men. The sea would save us with huge crops of high-protein foods and enough minerals to last an eternity. Science would develop ways of utilizing the life-saving properties of the sea, to keep our lives safe and comfortable.

All of which could be true, at least to some extent, except that a number of circumstances have interrupted, and are interrupting, the technological evolution necessary for this development.

We engage in emasculating wars, which not only use up vast

stores of natural resources but disallow realistic budgets for the development of oceanography.

The political jockeying for leadership in the aerospace program shrinks the budget of the economically more vital "inner space" program.

We have squandered time and money in all areas, except the one in which it would do the most good. Economic wealth does exist in the sea but we must work realistically to get at it.

We have already contaminated the air atmosphere. Now we appear to be launched on a superambitious program to contaminate the waters of the world.

Imbalance becomes a key word in our world: imbalance of our national economy, imbalance of our natural resources, imbalance of our atmosphere.

Fortunately scientists, engineers, and ecologists have begun work in areas that will tend to mitigate the situation.

Private industry has launched a very active oceanographic program, particularly in the areas of petroleum and submersibles.

The United States Navy, although understaffed and certainly underbudgeted, is accomplishing small miracles with its Sealab and affiliated programs.

Various departments of the federal government, such as the Bureau of Mines, the agencies concerned with fisheries, the Department of Health, Education and Welfare, the Department of the Interior, and other related agencies, are pressing for more adequate budgets and slowly getting them for developing the sea's resources.

But the budgets are still too small, and consequently progress is too slow. If we could have used but a small percentage for oceanography of the money spent annually on the Vietnam war, or a small percentage of our foreign "give away" program, this country would be headed less toward bankruptcy and more toward a safer, more healthful, even more affluent society.

So far the largest slice of federal funds for engineering is being spent in the area of defense and national security. It has been that way for years and undoubtedly will continue. The federal government must, however, take a closer look at the overall situation and reevaluate overall needs. Hopefully, the result will be an expanded budget for ocean engineering, for without it we cannot make substantial progress in this, the most provocative of all economic areas.

Private industry is doing much for the development of the

technological "hardware" necessary for the exploitation of the sea's resources. To the layman many of these devices have a futuristic look, like something out of Buck Rogers. But they are, in reality, the results of engineering designed to accomplish a specific under-water task—a task highly necessary if we are to keep apace with the demands of an ever demanding, ever needful civilization.

EG&G International has equipped practically all the deep sub-mersibles with superior underwater cameras and light sources. The resulting pictures have greatly increased man's knowledge of the benthonic (deep oceans) regions, monitored the condition of trans-atlantic cables, and helped in locating the wreckage of lost sub-marines.

Doctor Harold Edgerton and his associates have also developed a remarkable array of seismic profiling and dual channel side scan sonar systems, which can perform with picturelike accuracy a large number of underwater chores, such as detailed bottom profiling, which reveals the exact topography of the bottom so that a cable road can be preselected before the cable is laid; salvage search for wrecked aircraft or vessels; archaeological search for sites hidden by sand or silt; relocation of oil drilling locations (acoustic re-entry); harbor surveys; and many other applications. While on an archaeological search off Khalkis, Greece, the experimental probe transducer was able to accomplish soft sediment penetrations to approximately forty-nine feet. Obviously this could lead the way to many presently unknown buried shipwrecks and sites of civilizations.

The Electric Boat Division of General Dynamics, among a host of engineering products, has two of special interest.

The first one has to do with completely equipped underwater oceanographic laboratories, which are to be termed "manned underwater stations." These scientific stations will be composed of two vertical cylinders, each 48 feet high and 30 feet wide. One of the cylinders will house a life support system. The other will be a four-deck housing and laboratory station, providing a comfortable environment in which five scientists will work at an ocean bottom depth of 6,000 feet for thirty days at a time. From here the work teams will be able to observe all facets of undersea activity either directly through viewports or through closed-circuit television. Instrumentation will be mounted on the outside. From a lower-level observation sphere the teams will be able to collect specimens of flora and fauna through the use of manipulator arms, which will bring them inside through the use of a lock in-lock out system. The other engineering advancement is in the area of manned submersibles. *Autec I* and *Autec II*—25-foot, three-man submersibles—are the latest developments in underwater craft having the capability of complete safety.

In an emergency, a lever is pulled that ejects the crew's sphere from the rest of the vehicle. The sphere then slowly rises to the surface and safety.

Under normal conditions, however, these highly maneuverable craft can perform a wide variety of chores through the use of highly tactile, almost human, mechanical arms which can grasp and operate a number of tools. The arm is manipulated by a "dry" operator within the craft.

IBM has designed and built a seagoing computer which has already served a successful hitch aboard a Scripps Institute of Oceanography research vessel. Even during heavy seas the IBM 1800 can and does perform four essential jobs:

1. Make routine log of marine environment, including measuring water depth, checking sea surface temperatures and salinity, calculating wind speed and direction, and taking air temperature and humidity measurements.

2. Collect data for specific scientific experiments: for example, to chart temperature and sound velocity and conductivity profiles or to calculate the biologically important properties of the sea from measuring instruments.

3. Work toward a more accurate knowledge of a ship's position at sea. With satellite navigation, the computer will be able to calculate the ship's position every ninety minutes as a satellite passes overhead and re-pilot the ship's course, if necessary.

4. Give the scientist a general-purpose computer with which to analyze previously unanticipated relationships from the reduced data. The scientist will be able to write completely new analysis programs or modify old ones while still at sea.

Corning Glass Works has produced a number of products that have proved themselves to be extremely valuable in oceanographic work.

Dr. William B. McLean, director of the U.S. Naval Ordnance Test station at China Lake, California, was one of the first to recognize glass as a potential material for submersible construction.

Although lighter than titanium or steel, glass has the inherent quality of becoming stronger under the pressures exerted by water depth. It also resists salt water corrosion and has the extreme advantage of being completely transparent.

Corning Glass has already built a glass-ceramic hull for use as a deep diving submersible, as well as a number of large hemispheres that provide completely transparent observation "platforms," on other types of submersibles. These spheres have, in testing, withstood pressures equivalent to more than 35,000 feet beneath the sea.

A number of smaller sphere and capsule products have been developed and used successfully as flotation for oceanographic instrument packages, as well as in antisubmarine warfare.

Four prototype 375-watt, 30-volt underwater lights, designed to withstand pressures, were produced for the Perry submarine-building company.

In a current and very ambitious program, Corning Glass and General Electric's Re-entry Systems Department are working on a research project to develop a sphere twelve feet in diameter that could house men and equipment at ocean depths of 12,000 feet.

The Sperry Gyroscope Company, a division of the Sperry Rand Corporation, long involved with marine products, has been developing an instrument and control subsystem for the prototype of what will undoubtedly be the submersible of the near future: the NR-1, a nuclear-powered, deep-diving, long submersible range, underwater research and ocean engineering vehicle for use by the U.S. Navy.

At an experimental site in Montauk, New York, Sperry engineers have been working with a rig designed to improve the quality of closed-circuit television photography.

Oceanics Division of the Interstate Electronics Corporation is one of the largest producers and designers of buoyed ocean instrumentation and data systems. The firm's buoyed instrumentation systems can be found at every level of the oceans from the surface down to 18,000 feet. These systems include remarkable sensors that provide vital information about every kind of ocean condition, such as the measurements of waves, current, temperature, turbidity, salinity, and pollution.

Electrodynamics Division of the Bendix Corporation has also developed a battery of efficient ocean instrumentation systems, such as the B-3c Roberts Meter, which can measure current speed and direction at any depth; various instruments for measuring waves and plankton; single-sideband sonar telephones (for Sealab III); transponders for accurate ship positioning and oil drill bit reentry. Recently this company developed, for the State of Alaska, a sonar rack coupled with hydrophones that counts accurately the number of salmon as they head upstream to spawn.

As we forage deeper and deeper into the sea, as we fly more and more passengers across the skies, the need for accurate, long-range weather prediction becomes more and more important. Private industry is doing its best to keep apace with this need through the development of ocean engineering—the design and production of ocean environmental instrumentation and telemetry. Surface currents, subcurrents, tidal movement, winds, water temperatures, are all ingredients that affect weather, and our improved understanding of these ingredients will provide capabilities for precise weather forecasting. The result for us is, of course, greater safety in the air, on the water, and on land.

Not only engineers, but physiologists, too, are helping to break down the environmental barriers of the sea.

Dr. Johannes Kylstra of Duke University has been conducting successful experiments in which mice have actually breathed liquid without succumbing to what should have been inevitable drowning. The mice were immersed in and actually (although laboriously) breathed a fluid called fluorocarbon, which is supercharged with oxygen—some thirty times as much as the amount contained in air.

If the use of breathing liquids, instead of inert gases, can be developed for man, he will be able to dive deeper and stay down longer than has ever been dreamed possible. Furthermore, he will be able to come directly to the surface, without decompression or fear of the bends, regardless of his length of stay in deep water.

The Atomic Energy Commission has developed a radioisotope-powered, heated wet suit for divers. It will allow a diver to remain in cold waters without loss of body heat and, subsequently, energy. Feasibility tests of the suit will be conducted during the Navy's Sealab III project.

With more and more men exploring and exploiting the sea, the crucial need has arisen for an efficient Deep Submergence Rescue Vehicle (DSRV). Such a vehicle has been built by the Lockheed Missiles and Space Company. It is fifty feet long and weighs only thirty tons, so it can be flown anywhere in the world to a submarine disaster site. Piloted by a crew of three, and utilizing a system of directional and hovering motors, it will position itself over the escape hatch of a stricken sub and, after mating, effect the rescue of the injured vessel's crew.

Two major accomplishments by man are expected by the mid-1970's. Man will have landed on the moon, but what is economically much more important, he will be living in underwater habitats up to thirty days at a time, at depths of 1,000 feet.

By that time he will have begun to effectively exploit the wealth of the oceans' natural resources.

In the science-fiction world of the very near future we may very well have both sea dwellers and land dwellers. They will look exactly alike except that the sea dwellers will have a small circle at their windpipe where a tiny hole has been cut. This will allow their lungs to be filled with a special solution containing supercharged oxygen, which would give them the capability to work even in very deep water with no concern about decompression. The surgery required is minor, and in no way will affect their breathing while topside.

That gives us a few years to pray that man will have the spirit and intelligence to utilize these rich new areas for the benefit of *all mankind!*

12 · Recreation in the Sea

The American writer Ambrose Bierce described the sea as "a great body of water occupying about two-thirds of the world made for man—who has no gills."

George Herbert said, "Praise the sea, but keep on land."

"Love the sea? I dote upon it—from the beach." So said Douglas Jerrold.

These were the supposed wise men of their age, yet they were cynics. When it came to the oceans of the world, their vivid imaginations were blocked out. They could see no potential. I wonder what their reactions would be if they could return to life and revisit the present-day world.

What would happen, for example, if as they picnicked on a sunny beach with a beautiful girl and a bottle of vintage wine, the music of their poetry beguiling the lady, suddenly out of the sea came the entire membership of the La Jolla diving club, complete with rubber suits, masks, flippers, and air tanks?

Today millions of people of all ages are enjoying the healthful pleasures of snorkeling on the transparent surface of the world's waters strapping tanks on their backs and exploring the magical world below.

Scuba equipment today is virtually foolproof as long as you stick to brand names (discussed later in this chapter) and take a short basic course in sport diving.

Diving equipment is not difficult to use, and you don't need Olympic ability to enjoy the unforgettable delights of exploring an underwater garden in the multishaped form of a coral reef. I am not, of course, suggesting that you should rush out, buy a complete line of diving equipment, and hurl yourself off the nearest pier. Naturally you should take sufficient and competent instruction before going underwater with a compressed-air tank. I am saying, however, that skin and scuba diving equipment is easier to use than you might think. Although you are about to enter an atmosphere still referred to as "hostile," please be assured that, thanks to the advancements of technology, it *is* getting friendlier and friendlier.

Over the past few years I have made periodic visits to various diving schools and YMCA's and have watched private instruction. In the main, the teaching is done intelligently, with some emphasis on the psychological phases.

I have also watched the "hero image" teachers commit flagrant wrongs in order to make themselves look like Captain Marvel. This is the wrong method and succeeds only in discouraging would-be recreational divers. Class entry rules should not be so stringent that only Sandows and other strong men need apply. Whether or not you are blessed with these rugged physical capabilities, you have the unalienable right to enjoy the sensations of visiting the spectacular underwater world of the Florida Keys, the Bahamas, and the West Indies.

When we were children we were told by our parents to have fun but "don't go near that water." A little later we were told we could swim, but "don't go out too far." Every time anyone (usually the villain or the hero's best friend) falls overboard in a Hollywood movie, a shark's caudal fin is cutting the surface within seconds and zeroing in on the hapless swimmer. Frightening stories of undersea monsters, such as the giant squid, have been another heirloom left to us by even our classical writers.

If you have an understanding of this, progress will be much more rapid. To those of my readers who have never tried diving, I can only advise—try it! Snorkeling (lying on the surface with mask, snorkel, and fins) is the initial and easiest phase of skin diving. It can be enjoyed, barring any serious physical defects, by anyone between the ages of nine and seventy. Snorkeling allows you to look down and take in the loveliness of a world that gives you a whole new concept of form, color, and motion. Indeed, it gives you a whole new concept of beauty itself. Don't think of the underwater scenery as being blurred and ominous. As soon as you learn to wear the simple face mask and look down, everything springs into crystal clarity and you experience something beyond your ability to imagine. Your natural apprehensions and inherent fears will soon dissolve. If you have progressed to the use of a tank, you possess the ticket of admission to become part of that new world, at least for a brief stay. You can get a closer look at everything. You can glide like a bird or hover like a helicopter.

Sitting comfortably on the ocean floor, the author photographs a spearfisherman off the coast of Sardinia.

The author feeds tiny reef fish by hand on a coral reef in the Red Sea.

Striped parrotfish are part of the clear and colorful underwater playground in the Bahamas. (*Bahamas News Bureau*)

Divers examine a giant sea plume off Key Largo, Florida.

You can take an underwater camera with you and record forever the truly exquisite artistry of your surroundings.

Enjoy it as a family, for it will bring you closer together with sights to remember, experiences to treasure.

If you are wondering what equipment to purchase (the initial cost for mask, snorkel, and fins is about $17) and where to go for best diving activity, buy a copy of *Skin Diver* magazine. You will find it an excellent, informative, exciting, and sophisticated chronicle of diving adventure.

Before we leave the subject of snorkeling, I should interject a slight word of caution.

In your early training you will be told that in order to dive under and stay down longer without the use of a tank, you should hyperventilate your lungs. This is done merely by breathing in and out several times in rapid succession in short breaths or "pants." This charges your lungs with oxygen and allows you to hold your breath for a longer period of time.

This is safe to do, providing you do not overdo it. Over hyperventilation can cause anoxia, resulting in unconsciousness. If you limit your hyperventilation to about a dozen quick breaths, you will increase your ability to hold your breath without danger of anoxia.

Take the additional precaution of not overextending your capabilities and you won't have a thing to worry about.

One word of caution on the subject of tank filling. If the air-fill station is an unfamiliar one to you, carefully examine the setup. If the filling valve is too near the exhaust, don't have your tank filled. Poison air can result.

Some of the better-known manufacturers of diving equipment are: Dacor Corporation, Healthways, Scuba Pro, Seamless Rubber, Sportsways, Inc., White Stag, Voit Rubber, and U. S. Divers. As a sports diver, beginner, or veteran, you can feel perfectly safe purchasing equipment that bears the name of any of the above mentioned.

Your choice of scuba tank can be made from a wide variety of types and prices, ranging from the small tank with a capacity of 53 cubic feet ($56) to the 72-cubic-foot deluxe *twin* tank pack ($177). The best tank for the average adult to start off with is the 72-cubic-foot galvanized tank with .the J reserve valve, at a cost of $90.

Regulators range from $45 to $95. A good average price for an excellent single-hose regulator is about $70.

The snorkel should be a simple J-shaped rubber type, which sells for $1.95.

Flippers range from $3.95 to $15.50. A good average price is about $12. Flippers should not be too heavy, since an extra-large, heavy flipper only tends to fatigue the diver. The flipper should provide a full slipper (no open heel) and the fit should be snug but not tight. Be certain to try the equipment on before the final purchase.

Masks start at $4.95 and go all the way up to the new wraparound mask at $16.95. The $4.95 model will satisfy you for a long time.

Rubber wet suits range from $47 to about $80. This is for full suits and only necessary for long dives in cold water. A light weight "shorty" type (short sleeves and legs) can be purchased for about $35.

Spear guns also run the gamut of price range, but unless you are going after the monsters, a standard two rubber gun will serve you well for the smaller, dinner-size variety of fish, and will only take $23 out of your wallet.

An excellent stainless steel knife with a 6-inch blade and a sheath will set you back about $7.50.

A weight belt with eight one-pound weights will cost about $13.25.

Once you go in for more serious scuba diving the problem of timing your dive, and subsequently your air supply, becomes increasingly more important, so an underwater watch becomes a necessary part of your equipment. Again, the price range is very wide—from $25 to $300. I have used the accurate and rugged Rolex for many years and find the somewhat expensive price a good investment.

The manufacture of diving equipment and accessories has become a vast multimillion-dollar business, and it is growing every day. More and more sophisticated equipment is being designed and developed. Following are some of the more exotic accouterments:

Inflatable boats with transparent bottoms, which carry two to twelve passengers and are propelled by powerful outboard motors.

A low-priced gold dredge priced for those who want to sluice the gold-bearing rivers.

Singer John Gary's self-propelling back pack, called the Aqua-Peller.

The Nautilette, which is a two-place "dry" miniature submarine, selling for less than you need to buy a car.

Johnson's Air-Buoy, a small self-contained floating air "station," which delivers fresh air into two full face masks at the end of 25-foot tethered hoses.

Slurp guns which allow the diver to catch fish for his aquarium. You merely point the gun at a reef fish that you have selected and pull the plunger; the suction "captures" the fish and deposits him in a small holding chamber.

An underwater tape recorder for the marine biologist or archaeologist to record on-the-spot observations.

A portable decompression chamber that is low-cost, lightweight, and can be stored aboard the small family cruiser, making possible deep dives for valuable treasure.

A scuba board to replace the surfboard, for the more venturesome diver. The board is twelve feet long, has a viewing port to show what's below, and a rack to tie equipment to.

Underwater tows that look like miniature torpedos, with handles for the diver. Battery energized and motor driven, they tow the driver through the water and conserve air supply.

Underwater television provided by small, compact cameras that can be towed through the water by mechanical means or carried by a diver-operator.

A new and revolutionary scuba tank called Cryogenic, which utilizes liquid air, may be on the market soon. This lightweight rig developed by Jim Woodbury will allow a diver to stay down in deep water for longer periods of time, which will mean adherence to decompression tables. The anticipated price will be approximately $300 plus $3.50 for a liquid-air refill.

French doctors Fructus and Sciari have developed a hood that will enable divers in trouble to rebreathe their own expired air. This self-rescue system is based on the fact that a diver who runs out of air still has enough in his lungs and mask to enable him to reach the surface.

Still in the area of recreation, let's look at some of the contributions made to the sport of diving by television and motion pictures.

There is one man for whom I have the greatest admiration,

although at this writing I have not met him. He has contributed more to the development of the sport of skin diving than any other man in the world. That man is Hungarian-born Ivan Tors, who began his underwater career and influenced millions to do the same, with the production of the most famous of all underwater television series, *Sea Hunt*. He followed this with *Flipper, Daktari,* and *Cowboy in Africa,* plus a long series of theatrical films, including *Hello Down There* and *Around the World under the Sea.*

Mr. Tors's brand of television and screen entertainment bears the rare stamp of high quality and integrity. For his photography he maintains an underwater studio in the Bahamas, although much of *Sea Hunt* was shot at Silver Springs, Florida. Although the crew and cast of his underwater films may be surrounded by grouper, snapper, sharks, barracuda, or rays, he does not allow the use of spear guns.

He has complete faith in his theories on kindness and affection to all animals. He has also learned, through the years, a considerable amount concerning the behavior of fish, including sharks.

There was a scene in *Thunderball* during which James Bond was to effect an escape through a swimming pool filled with tiger sharks. Understandably, actor Sean Connery hesitated at the edge of the pool, staring with some apprehension as the dorsal fins cruised the water below him. Mr. Tors saved the moment by quickly stripping to his shorts and diving into the pool. After a few moments he stood dripping at the side of the pool, explaining that captured sharks are in a state of shock. As long as they are well fed they are lethargic and will not attack.

This is a theory which has been proved to me a number of times. Over the years I have swum in a number of pens full of captured sharks with never an accident. On occasion I have taken short rides on the sharks by grasping their pectoral fins. This was done not for sensationalism or because of the so-called "death wish," but rather to prove the theory that marine animals are not as dangerous as their reputations. In my first book, *Occupation: Adventure,* there are pictures and text describing the sensation one feels when riding a shark. The experiments were made in the hope that lives might be saved by lessening the chance of panic in the water—the fear of the unknown—which more than any other situation is a cause of drowning.

Although estimates vary, I would venture to say that there are at least fifteen million people today who don basic diving gear and snorkel wherever they can find a body of water. Not only is it a healthful activity, but it has resulted in a burgeoning industry employing thousands of men and women and contributing millions to our economy.

There is one encroaching danger, however, that unless controlled will threaten the life of this healthful American—or perhaps I should say worldwide—pursuit. That danger is pollution. We have already polluted the air we breathe and now we appear to be systematically polluting the waters of the world. This drastically affects not only the skin-diving population of our waters, but the fish population as well.

The pollution is the result of a number of man's unreasoning activities: the discharge of oil contaminants by ships at sea; the underwater and above-water detonation of A-bombs and H-bombs and other weapons; the abundant discharge of chemicals and sewage. The result is that many water areas throughout the world are unsafe for skin diving or swimming. The contaminated fish will often migrate to other areas, furthering and extending the contamination. Ultimately, these fish are consumed by human beings.

The danger is serious, but with stringent and scientific control it is possible to have certain waste effluents work in our favor. The warming of water in the vicinity of atomic power plants can help the fish population, especially bivalves, providing the water is not superheated and provided impurities, such as oxides and sulfur, have been removed. Certain waste products, such as domestic sewage, heated water, and carbon dioxide gas, can be utilized in a combined algal and shellfish culture system to produce a useful product while substantially reducing the effects of the waste.

The Electric Boat Division of General Dynamics has published a complete study that describes this potential in aquaculture.

I do not mean to minimize the danger of pollution. We urgently need to control the pouring of chemical wastes into our waters as well as to pass stringent laws forbidding the disposal of oil by ships at sea. The detonation of atomic and hydrogen weapons in air and water atmospheres should, for the sake of a heathful humanity, be outlawed altogether.

APPENDIX · *Recommended Diving Locations Around the World*

Interesting diving locations are limitless, and a discussion of them could easily make an entire book unto itself. There is fascination beneath practically every body of water in the world, whether it be a lake, stream, river, or ocean. It depends entirely on what the diver is seeking. Yet there is a common interest for just about everyone—a satisfying of curiosity, a search for adventure.

This chapter will take you to many diving locations throughout the world, from the most glamorous to the most rugged, to the almost prosaic. The diving locations are presented geographically. The listing starts with the United States, proceeds south to Mexico and the Caribbean, and then moves eastward across Europe to the Pacific Ocean.

United States

FLORIDA

The best diving locations are in the Florida Keys, with the exception of two freshwater rivers in the north central part of the state.

Both Crystal River and the lesser-known but equally beautiful Rainbow Springs are clear rivers where one can dive 365 days a year. You will find the bottom to be completely fascinating and totally different from what you have been experiencing in ocean reef waters. The flora is delicate and lushly green, and every shadow and shelf is alive with fish. Crystal River has one ingredient that may be considered a plus. You can get cozy with the blue-eyed, languid-lidded manatee as they glide and flirt enticingly nearby.

At the time of this writing Rainbow Springs is in the process of being transformed into a most beautiful and imaginative park complete with a monorail, animal reserve, aviary, quarter horse farm, a stern wheeler, fleet of submarine viewing-boats, and a scuba college.

Both places have comfortable accommodations, which should fit nearly any budget.

When exploring the Keys you will find yourself back in the exciting world of coral jungles. Striped grunt, blue runner, and coral-nibbling parrot fish swim through the reaching branches of staghorn coral. Squat, big-mouthed grouper frown as they hide in their caves. Slim, silver, torpedo-shaped barracuda flick their tails disdainfully in front of your mask or turn to bare their teeth. Sting rays swim by like graceful swallows. Tiny reef fish dart and swirl, making a fanciful kaleidoscope. Pink and purple sea anemones flaunt their spectacular beauty. Sea fans dance their weaving, undulating ballet. Moray eels, evil and lurking, wait impatiently for their prey. A small, almost transparent octopus tattoos himself against the dark coral rock. A queen conch shell blushes pink and pretty against the white sandy bottom. A red squirrelfish stares with big dark eyes from under a shelf. A patch of sea cucumbers forms a vegetable garden. A shark glides in silently. For a terrified moment everything freezes. Then all is panic. Everything flees, escaping!

The upper part of the Keys has been set aside as an underwater national park, so no spearfishing is allowed. The waters, however, are well worth visiting, and you have a large choice of accommodations.

From about the middle of the Keys all the way to Key West, which is the southernmost city in the United States, the spearfishing ban is not in effect, and you will find the underwater hunting excellent throughout. Bethal Shoals is the place to go for the really big ones!

In much of the area you have a choice of exploring the Gulf of Mexico or the Atlantic Ocean. For overall satisfaction I would recommend that you stick to the ocean side.

Although it gets rather warm, it is best to confine your diving to the summer months, when there are no storms to stir up the bottom. You will find the water clear and comfortable, and you will get plenty of action.

One last bit of advice. Keep your eyes peeled for ballast stones. They are small round stones used to ballast the old sailing ships. They indicate the presence of an ancient underwater wreck. A good find could pay for your entire vacation, even if you do have to give the State of Florida one quarter of the treasure cask.

If you have a few extra dollars, while you are in Marathon (Key Vaca), try to share expenses with a few others and charter a trip to Cay Sal in the Bahamas. It is just sixty-three miles away, and you will find yourself in one of the most interesting areas you have ever seen. Water with seemingly limitless visibility, big fish, underwater wrecks, huge lobster, uninhabited islands, dramatic marine topography, and sheer drop-offs into the purple mists.

Some years ago, on the Fresh Water Key, I found the date "July 1824" inscribed into the limestone rock, accompanied by a magnetic direction and an arrow. It may have been a message left by pirates, so try your luck at finding the buried treasure while you are there.

A word of caution. If you plan on making a landfall on one of the islands, remember they are British possessions and permission must be obtained. That, however, is a simple procedure and your charter boat captain will know how to do it.

INLAND LAKES

I must admit that I am not a lake-diving buff, but this type of activity makes the sport infinitely more accessible to many more people, and it is not without its inherent interest and pleasures. Following is a list of recommended inland lake diving locations:

Door County is on a peninsula located in the northwestern part of Lake Michigan at Green Bay, Wisconsin. The water here is fairly clear (about thirty-foot visibility) but quite cold, so take along your complete wet suit. The area is noteworthy because once there you can acquire a chart showing the locations of some two hundred underwater wrecks. Surrounding villages are quaint and historic, dating back a hundred years or more.

Upper Lake Huron and Georgian Bay are the locations of the clearest waters among the Great Lakes. Try the area off the town of Tobermory. Not only do complete diving, lodging, and eating facilities exist here, but there are some interesting wrecks. There are two old hulls within easy swimming distance of shore, the *Grand Rapids* and the *Sweepstakes*. For more excitement, you can charter a small boat and run out to Russell Island, just two miles away. Just offshore of the island is a stockpile of old wrecks, which has made this a favorite location among divers from all over the United States.

North and South Manitou Islands are located just off the northwest coast of Michigan. The water is clear, shallow, and not too cold in midsummer. Again, this location is for the disciples of underwater wreck exploration. When you find a wreck, examine it even if it has the appearance of having been thoroughly picked over. Constantly shifting sand may reveal a find where none existed the week before.

In New York State, Lake George and Lake Champlain are deep, clear, cold lakes with potential for exploration of historic shipwrecks.

CALIFORNIA

Diving practically anywhere along the California coast is interesting. Even the tide pools are alive with the miniature life of the sea, and you can while away many hours in total fascination. Don't just splash barefooted through these rockbound pools. Lie down at the edges and stare into the tiny bodies of water. As your eye adjusts, you will find myriad life there.

Never enter California kelp beds without a knife to cut yourself loose should you become entangled. That almost literally means never enter California waters without a knife because the kelp forests are everywhere offshore.

The average visibility, definitely less then in Florida or Bahamas waters, is perhaps thirty feet, although the difference in scenery makes it all worth while.

Look for rocky outcroppings along the shoreline. Most often where you find this condition, you will find interesting waters offshore.

I personally like the waters off Catalina Island best, especially the rocks just offshore, where good-sized sea bass are most apt to be located.

Between Laguna Beach and La Jolla are a series of small protected bays and inlets that provide interesting diving close to shore. You will sight plenty of fish, abalone, and very probably a few octopuses.

Equipment rental and air-fill stations can be found throughout the route.

If you get as far south as San Diego be sure to visit the famous diving club called "The Bottom Scrapers."

Regardless of where you are in California, have your wet suit handy (at least a shorty) because the water is quite cold.

Mexico

Diving is especially good in such areas as the islands of Cozumel, Isla de Mujeres, Medio, Lobo, and others on the Caribbean side. On the Pacific side try Tuxpan, Zihuatamejo, and even Acapulco. The waters are usually quite clear and cool, but the real excitement of exploring these places is the mystery—you never know what you are going to find. Anything is possible, from an ancient Mayan ruin to a sunken Spanish galleon. Check with the tourist office before you go, since some of the areas may not have the accommodations you would prefer. Most of them, however, have enough comfortable accommodations to satisfy the most discriminating traveler.

Incidentally, while in Mexico, if time permits, give yourself the rare and most individual sensation of diving in some of the wells. You may only run into a family of rare manatee (sea cows) or, on the other hand, it is entirely possible that you could discover a trove of Mayan sacrificial artifacts such as was found in the well at Chichén Itzá and valued at four million dollars.

Other good possibilities for treasure, in this case Aztec, are located in some of the high-altitude lakes, such as the Lakes of the Sun and the Moon, situated in the volcanic crater of Toluca. It is possible that one of these lakes holds the fabled treasure of Montezuma.

One of the best indications that the lake or well in which you find yourself diving may hold treasure is the presence of copal. This is a lightweight material that was used as incense during various ceremonial events.

Naturally, I assume none of my readers would ever attempt diving without the immediate presence of a buddy diver. There are other dangers inherent to diving in high-altitude lakes. The most important one is the lack of oxygen (up to 25 percent less than normal), which can lead to serious trouble if you overexert. You will feel comfortable underwater while breathing the air from your tank, but will find yourself gasping for air as you hit the surface and snorkel back to the shore.

Another thing to remember is that the bottom sediment can be easily stirred up, resulting in a complete loss of visibility in the areas surrounding you. The stirring up of the bottom is an occupational necessity, since it in only *under* the sediment that you may come on that fabulous treasure.

An additional reward for finding the treasure is that your wife will (may) forgive you for putting her through the many discomforts of such a trip. This is not for the city girl, whereas even Zaza would be happy at Cozumel.

There are other areas along the Gulf of California that furnish abundant activity for both the ardent diver and his family.

If you look at the map, you can easily see that the Gulf of California is a natural fish trap, and so provides a spear fisherman's paradise. A wide variety of sea shells can be found by the beachcombing family.

Guaymas is an excellent resort area that offers everything from luxury accommodations to complete skin-diving services, and it is serviced by Aeronaves airlines. The waters are alive with fish, and you may be treated to the rare experience of suddenly finding yourself swimming along with a 2,000-pound (but harmless) manta ray for a companion.

Mazatlán is the Mexico City of the California Gulf. It teams with life, and everyone in the family can choose from a variety of activity, all Mexican-flavored. The diving members of the clan will find an atmosphere fraught with flora and fauna aplenty. It is best not to plan your trip to Mazatlán during the rainy season, from June to August, because the summer squalls becloud the water. Fall and winter are the best times to plan your diving sabbatical.

The less said about Puerto Vallarta the better, for there a strange situation exists. Recently it was stung by the luxury stinger of Elizabeth Taylor, Richard Burton, movie company, *et al.*, and in its panicky reaction became neither fish nor fowl. Today it is neither authentic resort nor quaint fishing village. The prices, however, are authentic Palm Beach.

Other excellent and unspoiled diving sites, but with rather meager accommodations, are San Esteban and Tiburón islands, and Topolobampo on the mainland.

A word of caution. Watch the changes of the tides throughout. They are deep and quite strong.

The Caribbean

GRAND CAYMAN

Passports are not required here for American or Canadian citizens, unless you intend to stay for more than six months. After you have been here for three days, you may very well begin to wish you had brought your passport.

The soft tropical weather, the hospitable, cordial people, the dazzling beaches, the vivid, fragrant flowers, and the coral gardens set in transparent blue waters make this island a utopia for the diver-tourist or the jaded traveler.

And the flavor of past pirates adds spice to the very air!

Coral reefs circle the island, creating calm lagoons and providing safety and home to a myriad of brightly colored reef fish. The same reefs tore vicious gaping holes in sailing vessels that ventured too close, sinking them with deadly malevolence, then embraced them until they too became part of the reef.

There are charts showing the position of more that three hundred wrecks. Cannons, cannon balls, coins, cutlery, and pewter litter the bottom, but they are covered by a thin veil of white coral sand. Stories of buried or sunken pirate treasure abound in folklore and modern narrative.

Approximately two hundred and twenty-five separate species of seashells have been found, forming a treasure of their own.

On the east coast, Gun Bay provides strong evidence that it was once a stronghold of many pirates, including Henry Morgan and the much maligned Blackbeard. There are more than twenty wrecks in this area, from 200-year-old British sailing ships to a World War II Liberty ship.

The large barrier reef and fringing reefs outside Red Bay and Frank Sound are in brilliantly clear water revealing teeming fish life and eleven wrecks.

The largest of the reefs spans most of the north coast, and here you have a choice of both shallow and deep water.

Seven Mile Beach is one of the best family spots, since there are a number of good coral heads in shallow water for beginners and a liberal sprinkling of beautiful seashells on the beach.

The accommodations are plentiful and comfortable, with a selection to fit all budgets.

JAMAICA

The next island south and east of Grand Cayman that offers warm, clear water and many excellent diving sites is Jamaica.

As on Grand Cayman, American citizens are not required to have passports unless the visit is expected to exceed six months. However, proof of citizenship is required, and a return air ticket to the United States.

Diving equipment is available in Kingston for rent or sale.

Accommodations are many and varied, with a corresponding price range.

For the diving fraternity, the best time of year to visit is during the summer months. The weather is relatively calm then, and except for Kingston Harbor, good diving sites can be found around the entire perimeter of the island. You will also be taking advantage of more reasonable summer rates.

The sunken city of Port Royal is there, of course, but work has already begun on a methodical archaeological project that may hamper your personal activities.

The north coast is subject to a late morning wind, which may result in choppy water with limited visibility. However, it is always calm in the early morning. Here are a few recommended locations: The Blue Hole, near Port Antonio, is a beautiful deep-water cave for experienced divers only; Eaton Hall is also for the more experienced diver and requires a boat for transport, but the trip is well worth it; Silver Sands, too, is for the more experienced, since the reef—and a beautiful one it is—is about eighty feet deep.

For the family outings encompassing areas for both the beginner and the veteran, try the following: Braco, Montego Bay, Negril, Discovery Bay, Sans Souci, and Dun's River Beach (for the picnic crowd).

BARBUDA

This island also rates high among my favorite spots in the Caribbean. There are two hotels on the island—one extremely expensive, the luxurious Coco Point Lodge, and one very inexpensive, which goes under the rather dismal name of the Government Rest House.

The Coco Point Lodge has a monopoly on diving equipment, as well as an air-fill compressor. These are for the hotel guests only and are not for sale or rent. You can, however, rent a boat near the Government Rest House, so if you bring your own equipment you are all set to visit what can only be described as a skin diver's Eden, a City Celestial!

The intricate multishaped reefs create this atmosphere. The beaches are uncluttered, wide, white, and made of a sand so finely pulverized it could almost serve as Johnson's baby powder. The water is so clear it appears to be transparent. Even from your rented boat you can stare down and see details of the live and lovely coral reefs.

It is a place for prize-winning photography and trophy-size game fish.

I might suggest that the spearfishing buff try Codrington Shoals and Dodington Bank.

When the photographer tires of the spectacular underwater scenery, he can take dramatic pictures of eighteenth-century wrecks.

SABA

I usually skip Haiti and Puerto Rico for a number of reasons. The main one is that the diving locations are not good enough to warrant the indifferent, often discourteous treatment meted out to the tourist. So, after a visit to the Virgin Islands, I stop at Saba.

This is a free port. Only proof of citizenship and a return air ticket are required.

As is usual for divers, the weather is best during the summer.

Hotel accommodations are somewhat sparse, but you should have no trouble during the summer. At last count, there was one hotel and four guest houses. I'm certain, however, that by now there are more accommodations, because this is a lovely, unique, unspoiled, and very picturesque island. It is bound to be discovered, and soon!

Except for the east coast, which suffers the brunt of the prevailing winds, there is virtually no place off the beautiful Saba shores where the diving is not ideal and totally absorbing. There are countless coral gardens and caves. Fish activity is a constant moving parade.

Although spearfishing is excellent throughout the island, I would recommend that you rent a small boat and have a guide take you out to the Saba Bank, where you will have a spearfishing experience you will not be able to forget for a long time.

Another plus is the presence of colonies of both green and hawksbill turtles. You do not have to worry about the giant green turtle, since he is a gentle vegetarian. Do not get too cozy with the irascible hawksbill, however. He is *not* a vegetarian and he can take a very painful bite of your arm or leg. Just leave him alone and he won't bother you at all. The shell of a hawksbill has a more humped look and his head and bill are longer than those of a green turtle.

ANGUILLA

Very much the same diving conditions and tourist accommodations exist here as at Saba, except that the latter is mountainous and Anguilla is low and flat. It possesses the most beautiful beaches in the Caribbean.

The water visibility is at least 100 feet, and this is truly virgin territory for both the spearfisherman and the underwater photographer. The coral reefs are matchless in their unspoiled beauty.

VIRGIN ISLANDS

The Virgin Islands, which include St. John, St. Thomas, and St. Croix, constitute the epitome of ideal diving locations. The water is warm and clear; lush coral reefs and fish life are abundant.

St. John is largely preserved as a national park and regulated by the U. S. Department of the Interior. It would be best to leave your spear gun in the hotel room and bring your underwater camera. The fines are stringent should you wander into park area, thinking you are in unregulated territory, and spear a fish. Almost three-quarters of the reefs and their inhabitants are enjoying the protection of rigidly enforced laws. Don't feel too disgruntled, because this does have its compensations. The fish all seem to have come to the comfortable realization that they are protected, so they do not "spook" into the nearest cave at your approach. The variety of fish is infinite, and they will swim right into your camera lens and pose on cue.

Some recommended locations on St. John are Johnson's Reef, Cinnamon Bay, Trunk Bay, Chocolate Hole, and Stevens Cay (site of wrecks of World War II gunboats).

Diving equipment facilities and lodging are very good throughout.

St. Thomas is the "Riviera" of the Virgin Islands. People from everywhere in the world come here, so this is where the action is. Sophisticated and polished, it has the best shops, hotels, and restaurants in the Islands and probably the most expensive taxi service in the world.

Complete diving equipment and charter boat facilities are available.

Spearfishing is permitted, but regulations have probably been enacted by now, so check with the island's Commissioner of Water Sports, Department of Commerce, before you plan your diving trips. Of course, if you charter a boat, the captain will be aware of any new regulations.

Some recommended diving locations are Stumpy Bay (for advanced divers), Coki Bay, Sapphire Beach (excellent for beginners), Cabrita Point, Brewers Bay, Flamingo Bay, Bluebeard's Beach, Morningstar Beach, Honeymoon Bay, Limestone Bay, and Shipwreck Point.

For the most experienced divers, I suggest a trip to such nearby islands as Thatch Cay, Inner and Outer Brass Islands, and Grass Cay. Fishing for the big ones can be done here, especially on the outer edges of the reefs, where the water depth drops off sharply.

St. Croix is the largest of the Virgins and the most anomalous. For a tropical island it has a most diversified complex of scenery, with everything from a lush rain forest to a cactus-spotted desert. There is fascination for everyone in the family.

The underwater reef will make even the most traveled, jaded photographer gasp with pleasure. Buck Island Reef was proclaimed a National Underwater Park Monument by President Kennedy in 1961. The length of this spectacular reef is four miles, and it offers an almost unending variety of caves, grottos, and flagrantly beautiful coral structures.

Reefs such as Barracuda Grounds, Porpoise Patch, and Middle Ground can be reached by swimming a short distance from shore. These are clear, shallow reefs that furnish excellent training grounds for the novice, before he attempts the larger reefs in open water.

Other reef recommendations are Cramer's Park, Christiansted Harbor, Pelican Cove, Cane Bay, Salt River Bay, Seven Mile Reef, Jack Bay, East End, and Turner Hole.

Although this island is somewhat off the tourist route, you will find comfortable accommodations and safe, efficient diving equipment service centers.

THE BAHAMAS

These islands begin fifty miles off the Florida coast and curve southeastward in a 750-mile arc. There are more than seven hundred of these British-owned jewels (some have made a bid to be autonomous), scattered over some seventy thousand square miles of sparkling sea.

The best time to visit is during the summer, when the water is warm and exceptionally clear. March is a good month to avoid altogether. The water is cold and murky.

Some of the islands that remain relatively unspoiled but have comfortable facilities are Andros, Exuma, Harbour Island, Spanish Wells, and Abaco.

You can snorkel off any of the white, wide beaches and find something interesting close to shore. After you have had enough of the shore area, check in at the closest diving headquarters and have someone who really knows the area take you to some of the truly spectacular outer reefs.

On *Andros*, the largest of the islands, the skin-diving headquarters is at Small Hope Bay. Visit a water location called Tongue of the Ocean for a once-in-a-lifetime thrill.

Exuma—which consists of two islands and various cays—is the undisputed beachcomber's paradise of the world. Most of its cays are uninhabited, nearly all have perfect anchorages, and the clear sparkling waters have strewn rare and exotic seashells along miles of golden, unpopulated beaches.

The diving headquarters is located on Staniel Cay.

Harbour Isand is a small cay, three and a half miles long, with broad beaches of powder-smooth pink sand. The pink color comes from the mixing of pulverized coral and conch shells with white beach sand.

St. George's Cay and **Spanish Wells** are located a scant ten miles northwest of Harbour Island. Spanish Wells has an ideal harbor for small boat and yacht anchorage, limitless white sandy beaches sloping gently into shimmering topaz water, big-game fishing, and a storybook world of underwater reefs and multihued fish.

One of the most complete diving centers, located on Spanish Wells, is called the Lloyds, after the family that owns and operates it. The Lloyds have everything in the way of equipment from a miniature submarine to flippers—plus a thorough knowledge of the surrounding waters.

Abaco is a most unusual island in that it possesses wide expanses of thick pine woods dotted with small lakes, which give it the appearance of a flat version of Switzerland.

Hope Town, one of the settlements, which looks like Cape Cod, has a gay 120-foot, barbershop-striped lighthouse marking its excellent harbor. There is a historical reason for the incongruous Cape Cod look. English loyalists fled to Abaco from the American colonies in the 1700's. Most of these loyalists remained on Abaco even after the end of the Revolutionary War. Throughout the villages you will hear eighteenth-century songs sung just as they were in colonial times, and you will meet blond, blue-eyed descendants of the loyalists.

Skin and scuba diving is matchless throughout the island's waters, and it holds an extra and exciting attraction. For anyone who wants to try his hand at treasure diving, there are isolated cays off the west coast where it might just turn out to be profitable. Not too long ago one fortunate diver, named Roscoe Thompson, found a Spanish silver ingot worth $20,000 in the waters off Gorda Cay.

As on the island of Andros, when you tire of diving you can arrange for a wild boar safari or head for the northern part of the island in the hope that you may catch sight of some of the small wild horses that still roam the area.

The French Riviera

Not only is it illegal to spearfish while using scuba equipment, but a license is necessary to spearfish at all. If you get to Paris you may obtain the required permit at the Confédération Mondiale des Activités Subaquatiques, 3 rue du Colisée.

There is an abundance of diving gear, filling stations and boat charters throughout all diving areas.

Recommended areas are Marseille (try Ile Maire), Le Grand Congloue, Bandol, St. Tropez, La Trayas (underwater Notre Dame Cathedral), Cap d'Antibes, Cap Ferrat (location of Cousteau's *Conshelf III*).

Although the diving at **Monaco** is usually not very satisfactory due to lack of good visibility and fish, a visit to the Oceanographic Museum is worthwhile and will flag your lagging spirits, should you have encountered a brief spell of bad weather.

Monaco, because it is a lovely, sophisticated city filled with lovely sophisticated people, will, as it did with me, get you out of the doghouse with your wife for spending too much time underwater. The *pièce de résistance* is the Gala Dinner, which is held every Friday evening complete with dinner, champagne, dancing, stage show, and fireworks.

As a matter of fact, I enjoyed it as much as my wife did. It was good to get dressed again, go to smart, interesting restaurants serving superb food and vintage wine, dance under the stars with my exquisite, glowing vision of a wife, and walk hand in hand window-shopping along the laughing summer streets.

I guarantee that the respite in Monaco will most certainly revive your lagging spirits, whatever they may be.

Malta

I do not know this group of islands as well as I know other parts of the Mediterranean, but I did visit it briefly in the summer of 1968 and can make the following comments.

The visibility of the water is generally excellent, averaging approximately one hundred feet during the summer months.

St. Thomas Bay and the adjacent area provide both snorkeling and deep-water diving areas, with a great deal of fish activity and interesting underwater scenery.

The west coast, especially around Golden Bay, provides a varied menu for all diving tastes and degrees of experience, and the accommodations are luxurious.

Gozo is another Maltese island that has very good living throughout and is well serviced by fast hydrofoil from Malta. The accommodations are excellent.

Comino is a small island that until very recently was undiscovered by the floodtide of tourism. Even today it is virtually unspoiled, but its popularity is growing. This island, too, is representative of the entire group, with rocky, indented coastline, clear water, plenty of game fish, ancient wrecks and amphoras.

As always, should you find amphoras, leave them at their sites, but report your finds to the local museum. You will be doing the country, history, and yourself a service.

Italy

You will find reasonably priced diving equipment of good quality throughout the diving centers. Except for certain of the islands, your underwater camera will get a better workout than your speargun.

On the mainland, the west coast has the only really worthwhile diving spots. Remember, the laws here strictly forbid taking archaeological artifacts from the bottom. Otherwise the country does all it can to make your underwater holiday a pleasurable one.

Recommended locations on the mainland are the Portofino peninsula, San Fruttuoso Bay (statue, "Christ of the Abyss," can be seen in about fifty feet of water), Punta del Mesco, Punto di Monte Grosso, Capo Cavallo (a few wrecks here), Capo d'Anzio (caves, wrecks, but hip-deep with umbrellas and people on the beach), Sorrento, Salerno, and finally Palinuro.

The offshore islands will give you an opportunity to test your skill with the speargun again, but keep the camera handy, for these are the areas that provide the best diving.

Isola di Capraia has exceptionally good diving conditions on the western side, where there is good protection against the wind. Living accommodations are quite sparse here, but this island does connect by ferry with historic Elba (also very good diving), where the accommodations are somewhat better.

Isola di Montechristo is the spot for stalking and spearing good-sized game fish, which can give you the kind of action you've been looking for. Before you leave Elba, you will have to get permission from the police to go ashore on that particular island. This is a rugged island with no room (not even in the manger) for the tourist. It is a camping-out situation, which is in itself a rewarding experience—but only for the stalwart.

Isola di Giannutri is another location well recommended for spearfishing, but again there are no tourist facilities. As a matter of fact, you had better return on the same boat on which you came, or you may find yourself stranded for a few days, since the boat excursions are sporadic.

Ponza and **Palmarola** tend to get a bit crowded, especially over the summer weekends with local divers—albeit the diving *is* interesting.

Capri, near Naples, is one of my favorite places, since there are interesting things to do both underwater and on the surface.

The ethereal beauty of the famed Blue Grotto, Emerald Grotto, Three Sisters Grotto, and other large caves, will provide a new dimension of excitement for the family diving excursion. Two badly eroded statues discovered in the Blue Grotto just a few years ago may date back to Emperor Tiberius. The ruins of his palace and that of Augustus Caesar can be explored on the island, as can other Roman ruins. Hire a boat and guide at Marina Grande, circle the island (nine miles), and get into the water wherever it looks most promising. The Quisisana is the most elegant hotel, but a complete range of accommodations are available.

Isole Eolie are a group of islands near the southernmost tip of Italy that provide good spearfishing throughout, with satisfactory tourist facilities.

Yugoslavia

You will require a visa to visit Yugoslavia, but it can be acquired easily at the point of entry.

It is necessary to register all diving equipment when you enter the country, and it must be in your possession when you leave.

A spearfishing permit is also required, and spearfishing while wearing scuba equipment is considered illegal. As a matter of fact, the use of scuba equipment is prohibited in a number of areas, so check with the police in the areas where you hope to dive.

The penalties for taking or destroying archaeological artifacts are severe: five years in prison plus a heavy fine. If, on the other hand, you make a hitherto unknown and important find, you may find yourself the recipient of a reward.

For my money, the best diving areas are in the southern part. Here the water is very clear and considerably warmer than elsewhere in the country, with fascinating bottom topography, ancient ruins, and a better-than-average fish population.

The diving areas from Dubrovnik south are comparable to the best underwater locations anywhere in the Mediterranean. While you are in the area, look at the ruins of Epidauros in Thia Bay, just a shard's throw from the Hotel Epidauros.

Greece

The only time to visit Greece is during the warm summer months, when the water is warm and clear. Then anyone can snorkel and stare down into the limpid, crystalline depths to his heart's content.

But strap a tank on your back and in all innocence dive in an area of known or suspected underwater antiquities, and you can find yourself in trouble. The fines and penalties are severe. Spearfishing is always prohibited while wearing scuba.

If you are planning a series of diving trips, and I can assure you such trips are eminently worthwhile, it would be best to apply to the Greek Federation of Sub-Aquatic Activities and Fishing Sports, Aghios Kosuias, Glyfada, Athens, Greece.

If you intend to pursue your diving activity more casually, check with the nearest Coast Guard station or Greek diving club and announce your intentions.

You will have no trouble finding diving equipment to buy or rent in Athens, but once away from the city you will find little if any equipment available.

Although my deepest love is for the islands, here are a few recommended places well worth exploring off the mainland: Gulf of Krissaios holds some interest. Close to the village of Itéa are the wrecks of a number of Turkish warships.

Cape Soúnion is a short car ride from Athens and does provide some interest, especially the island of Makrónisos. As a dry side trip, be certain to visit the ancient Temple of Poseidon.

Platania is not only a quaint village, but the rocky shoreline from here northwest to the town of Agiokambos provides interesting spearfishing (snorkel only).

Peloponnesos, still close to Athens, is separated from the mainland by the Corinth Canal. There is a great deal of good diving throughout the area, but here are a few of the choice spots: Navarin Bay contains more than forty Turkish ships sunk in the early 1800's. They lie in rather deep water, from 70 to 125 feet; although this necessitates a brief visit (because of depth), it will be an interesting one.

Plitra is a small village on the eastern shore of the Gulf of Lakonia, and just offshore its town pier can be seen the remains of a submerged village.

The area around the village of Monemvasía is wild and rocky, providing an excellent area for spearfishing.

THE GREEK ISLANDS

Hydra is a short hydrofoil boat ride from the mainland, and although it is becoming somewhat commercial it still retains some of the look and flavor of the old sponge fisherman's village it once was. Rock caves, crevices, and clear warm water make the island well suited for diving.

Corfu, located on the Ionian Sea near Communist Albania, is a very popular summer resort with all types of accommodations. The best area for diving is the west coast, which can easily be reached by road or a rented fishing boat.

Rhodes is located just off the coast of Turkey and is not only a beautiful island but provides interesting dry side trips to various ruins. The Cape Kopria area may be considered the better diving area, but the entire coastline is excellent.

Melos lies between Piraeus and Crete and can be reached only by island steamer from the mainland. The island itself is in the beginning stages of being "discovered." The diving everywhere is superb, with all manner of past history to beguile you. The dry excursions will reveal early Christian catacombs, an outdoor amphitheater, an acropolis, old Roman baths, a beautiful Greek church, and a museum of antiquities. The famous statue of Venus de Milo was discovered here, as well as that of Apollo. Among other artifacts in the museum, you will see the ancient bowl of brass.

The waters are a treasury of caves, amphoras, ancient wrecks, and sunken cities. Look, but do not take! Resist the temptation to pick up a partially buried amphora, because by so doing you may be destroying forever a major historical site.

There are a few small hotels on the island now, so your entire family will be perfectly comfortable and you will all enjoy days of unforgettable activity.

Crete is another favorite of mine in the Cyclades. It can be reached by ship or air from Athens, Piraeus, and Rhodes. One cannot be aware only of diving while on this most fascinating of islands. The ruins, dating all the way back to Minoan times (3000 to 1250 B.C.), especially the palace of Knossos, will hold you spellbound as you wander through its labyrinth. Stories that for eons were thought to be only myths are given startling credibility in the light of recent archaeological discoveries. All the glories and splendors of the past are coming to light, and they are there on the island of Crete for you to see. To help you enjoy it all to its fullest, read up on the history of Crete before you leave home.

On the north coast good diving depends on the abatement of strong northerly winds. When it is calm, and that is rare, the diving is excellent. Avoid Kismou Bay, where the bottom is only sand, and uninteresting. Souda Bay is the most protected area on the north coast and so offers the best diving.

All the other coastal waters, south, east, and west, offer nearly ideal diving conditions. Do a lot of snorkeling until you find something of particular interest that invites further investigation, whether it be rocks, wrecks, or underwater caves.

Two areas of exceptional interest are Matala Bay, where Homer's *Odyssey* reports the sinking of ships belonging to the king of Sparta, and a stretch of about twenty-four miles between Points Flomes and Mouros.

One property inherent to all the waters around the Greek islands is a quick drop from shallow to deep water, but this need not bother you. Never stay in deep waters long enough to necessitate the use of decompression tables. Remember, you won't find a decompression chamber on a Greek island. And, of course, never dive alone, no matter how tempting the situation.

Turkey

Turkey is considerably behind Greece when it comes to the development of underwater sports, although it is at least comparable when it comes to the quality of diving locations. There is a dearth of diving equipment, and what little equipment is available is very expensive. The best thing is to have your own equipment, including all spare parts. To curtail expense, the heavy equipment can be shipped by boat well ahead of time, or as air freight or unaccompanied baggage.

As in Greece, the removal of any submerged antiquity is strictly forbidden by law. The same applies to surface antiquities.

Although Turkey possesses four coastlines, we will only be concerned with three—the Mediterranean, the Aegean, and the coast of the Sea of Marmara. The turbidity of the Black Sea offers no enticements to the far-traveling diver, especially compared to exceptional clarity of the other locations.

The Mediterranean coastline is fraught with the compelling stories of mythology as well as numerous archaeological sites. The ancient city of Olympus at Cirali Bay, the burning mountain of Chimaera, and the thousand-year-old Roman city of Phaselis, are just a few. While at Phaselis break out your snorkel equipment and take a look at the sunken walls and pier just offshore.

If you stay away from the flat gravel coastline, which is uninteresting, you will find good diving around the rocks and caves near Antalya.

There are also underwater ruins near the small town of Mersin, which once was a part of the once important city of Pompéiopolis.

For a change of pace, look for large grouper in the areas of rocky coastline. A day or two of spearfishing here breaks up the doldrums and renews interest in archaeology.

The Sea of Marmara is technically a large inland sea, connected to both the Black Sea and the Aegean. The visibility is good, averaging about seventy-five feet, but considerably less than in either the Aegean or the Mediterranean, which offer visibility in excess of a hundred feet.

The best areas for diving appear to be the islands, such as Marmara, Princes, Yassi, Büyük, and tiny adjacent islands.

The bottom is fairly interesting, with rocky patches and good fish activity. Frequently you may run into a school of dolphin, who are just as playful as the variety found in American waters. Halibut and lobster are also rather plentiful.

Avoid the mouths of the rivers and the Aegean will offer the most magnificent diving in all of Turkey. So many ships plied these waters that the bottom is liberally sprinkled with amphoras. The presence of amphoras does not necessarily mark the site of a wreck. It could merely mark the path of a trade route or perhaps an anchorage. Today an anchorage litters the bottom with Coke bottles and beer cans. In antiquity it was amphoras and anchor blocks.

When budgeting your time, allot the major portion for this phase of your travels, because not only are the diving locations numerous, but some of the archaeological sites are "musts." Troy, for example, is fascinating to visit. It requires a visitor's permit, which may be obtained in the city of Canakkale. Other sites include Alexandria Troas, Izmir (ancient Smyrna), Lebedos (site of ancient Greek drama), Ephesus, Didyma (site of Temple of Apollo), Bodrum, Patara, Kos, and Kim Point (Byzantine city of Andriace).

A complete adventure in diving can be found in the area from the southern shore of Izmir Bay to the entrance of Kuşadasi Bay. You can hire a boat for little money in Izmir and look for shallow rocky reefs. Not only will you find large schools of fish, but you stand a good chance of coming across an ancient wreck.

A rocky peninsula between Kizil Point and Kerme Bay offers perfect diving sites throughout. When you get to the town of Bodrum be sure to visit the Museum of Objects Found in the Sea, which is run by an American friend of mine named Peter Throckmorton. Drop in and give him my best wishes.

Boz Point in Hisaronu Bay not only has good spearfishing, but may allow you to discover underwater ruins. History tells us that this was the site of a well-populated settlement.

Another good bet for photographing and exploring underwater ruins is the area around the town of Fathiye, especially at nearby Tersane Island.

You should stop at the village of Kaz to see the ancient Greek theater. Approximately three miles east is the island of Kelova, where a small bay named Ucagiz is situated. Ask to be directed to the ruins of a castle where, in quite shallow water just offshore, you can see sunken sarcophagi scattered over the bottom.

If, after the sights of Greece and Turkey, you can sit in Spartan silence, I have a few more magical places to beguile you with.

Cyprus

While on Cyprus it is best to go back to the archaeological digs and forget about spearfishing for a while. The island, which has known a succession of civilizations for 8,000 years, is a historical thesaurus of ruins, both on the surface and underwater. The island is also interlaced with a network of roads that can take you everywhere, so the best way to see it and visit the diving sites is by rented (very reasonable) car. There is an abundance of hotels to fit all budgets, and a diving equipment center at Famagusta.

Cape Kormakiti is a good place to try for ancient underwater wrecks.

Salamis is a worthwhile location to explore both for underwater ruins and wrecks.

Cape Dolos, on the south coast, is usually good for both spearfishing and wreck exploration.

On the east coast Famagusta has lovely beaches and excellent underwater hunting.

My favorite spot on all of Cyprus is the area around the Akrotiri Peninsula. I have taken a number of sizeable grouper from the rocks underneath the cliffs, as well as photographed evidence of ancient wrecks. Besides the deep water, there is a considerable area for family snorkeling fun.

Incidentally, it is wise to check with the police before you begin your junket, because from time to time certain areas are declared off limits for tourists.

Israel

Although this country does not have as much coastline as most other Mediterranean countries, the diving locations more than make up for it in archaeological treasure. Israel has an added diving plus in that within its boundaries are two famous and historic bodies of water—the Dead Sea and the Red Sea.

The Dead Sea, mysterious in the thick veil of a timeless historic past, has a diving center at Ein Bokek on its southernmost end.

The Gulf of Aqaba, the northern arm of the Red Sea, is also within this country's borders. In my experience, the Red Sea has the clearest water, the most luxuriant coral reefs, the largest fish population, and the most sharks of any comparable body of water. The diving center there is in the town of Eilat on the northwest tip.

On the Mediterranean coast there are interesting, easy-to-reach locations off Akhziv, Acre, Haifa, and Tel Aviv, but the most complete and impressive area is at Caesarea, whose seaport was built in 25 B.C., as a costly tribute to Caesar Augustus.

A tragic series of misfortunes befell this magnificent city-seaport: A geological fault and several earthquakes, plus invasions by Arabs and the Crusaders, erased all but the faintest signs of its former glory. Settlements were built over the remains, statues were fragmented into material for roads, columns were cut up into milling bowls, sand swept over and covered some of the constructional skeletons, and finally the remains of the city sank wearily into the sea.

As late as 1957 Baron Edmond de Rothschild, seeing its potential, began to create a beautiful tourist resort out of the withered bones of the past. Modern hotels and ancient restorations appeared back to back and became neighbors.

The beaches are unspoiled, since their access is only by road. Diving is a day-to-day adventure. At Caesarea you merely follow the trail of amphora shards into the water. You can still see the ruins of the 2,000-year-old Roman harbor—columns and amphora shards litter the water.

There is no point in taking along your own tanks and regulators. The price of good rental equipment is quite reasonable.

Incidentally, don't expect exceptional visibility in the harbor area. During the summer months it will average only about forty feet, but it is enough to spot a half-buried amphora. The waters of the surrounding areas are, of course, much better, with visibility of eighty feet and more.

French Polynesia

Eight hours by jet plane will land you amid the French Polynesian "flower people." They look, think, wear, and live flowers of the most miraculous colors. The air is redolent with their perfume. The water is an indescribable blue.

Bora Bora is not only one of the most beautiful islands in the world, but its lagoon is one of the loveliest. It is protected by a fringe reef offshore, making for ideal family snorkeling conditions. The water is extremely clear and the thick coral reefs support a constant parade of tropical fish, cowries, and sea anemones. The reefs are mostly shallow and sun pouring down gives them a golden aura. You can safely add to your rare shell collection, but beware of the soft *pink* anemone. Its tentacles will sting painfully and leave welts that last for days.

The outer reef is even more exciting, with larger, more luxuriant formations. In the distance you will probably spot giant tuna and an occasional shark. In the deep ravines you will brush up against huge grouper, jack cravelle, and schools of barracuda.

There are more than a hundred and twenty-five islands in the Polynesian group, but one that every clear-thinking diving family should visit is Woorea, where the Club Méditerranée is situated. Club members live in a thatched-roof village of some eighty huts. The dress is totally informal, with the ladies wearing bikinis and the men wearing the pareo, which is a kind of mini-sarong.

Diving in both the lagoon and the fringe reef is superb, but be prudent about watching the tidal changes. The water will move rapidly through the open channels in the reef.

All in all it's a great big wonderful world, with beauty and humor everywhere.

Travel wisely and enjoy it all.

I never will stop traveling, so may I close this chapter with the sincere wish that we may run into each other one day and share an adventure together! Just introduce yourself and we will be on our way.

Index

SALVAGE